New Comparisons in World Literature

Series Editors
Upamanyu Pablo Mukherjee, Faculty of English, University of Oxford,
Oxford, UK
Sharae Deckard, School of English, Drama & Film Studies, University
College Dublin, Dublin, Ireland

New Comparisons in World Literature offers a fresh perspective on one of the most exciting current debates in humanities by approaching 'world literature' not in terms of particular kinds of reading but as a particular kind of writing. We take 'world literature' to be that body of writing that registers in various ways, at the levels of form and content, the historical experience of capitalist modernity. We aim to publish works that take up the challenge of understanding how literature registers both the global extension of 'modern' social forms and relations and the peculiar new modes of existence and experience that are engendered as a result. Our particular interest lies in studies that analyse the registration of this decisive historical process in literary consciousness and affect. We welcome proposals for monographs, edited collections and Palgrave Pivots (short works of 25,000-50,000 words).

Johan Höglund

The American Climate Emergency Narrative

Origins, Developments and Imaginary Futures

Johan Höglund
Department of Languages
Linnaeus University
Kalmar and Växjö, Sweden

ISSN 2634-6095   ISSN 2634-6109 (electronic)
New Comparisons in World Literature
ISBN 978-3-031-60644-1   ISBN 978-3-031-60645-8 (eBook)
https://doi.org/10.1007/978-3-031-60645-8

Cover credit: Terry Moore/Stocktrek Images

This Palgrave Macmillan imprint is published by the registered company Springer Nature Switzerland AG
The registered company address is: Gewerbestrasse 11, 6330 Cham, Switzerland

If disposing of this product, please recycle the paper.

This book is dedicated to my parents for all their love and support.

PREFACE

This is a book about fiction that insists that climate change is an emergency not for the planet but for American capitalist modernity. Most of the texts I discuss are informed by the notion that humanity as a species has caused this emergency, and they call upon the militarized security apparatus of the US to resolve the present crisis. In the works studied, the lives of white, middle-class people are suddenly and dramatically interrupted by the detrimental effects of climate change. These effects consist of the sudden influx of migrants, other nation-states competing for increasingly scarce resources, or even the planet itself, rising as an angry, chthonic monster. Such depictions, the book argues, transform the attempt to prevent global warming into an armed battle between a damaged planet and the institutions and mechanisms designed to secure capitalist modernity. If this battle is won, American hegemony is reinstated and business as usual resumes. If it is not, the world ends.

Much of the fiction that I discuss has been classified as climate fiction, but in this book, I recategorize the texts I discuss as *American Climate Emergency Narratives*. I also show how this type of fiction emerges out of a long tradition of American writing that can be traced back to the early colonial period when settlers arrived on American Indigenous land and began transforming it into plantations, mines, paved roads, industries, airports, megacities, and military stations. As a growing body of environmental history and climate research reveals, this arrival and the transformation that followed are the actual beginning of the biospheric crisis. However, in the stories settlers told themselves, and told the world, their arrival to America, and their colonization of much of the continent, are described as the dawn of civilization.

Growing up, I was surrounded by such American stories. They insisted that I must love but also police my white, male body, that I must be straight, disregard the work done by women, disrespect Indigeneity, queer people, and persons of colour, think of land as property and resource, and always invest in my own entry into capitalist society. My conception of the world, people, and the planet has been indelibly affected by this culture. Since I began researching literature, films, and games, much of my scholarship has been about these very texts. Fortunately, it was not the only story I was told. I vividly remember reading Ursula LeGuin's *The Word for World is Forest* (1972) as a young teenager and how weirdly it gelled with the stories told, and the activism performed, by the environmental movement of the late 1970s. Speculative, feminist, radical, and postcolonial fiction from all parts of the world offered a vantage that made it possible to understand and critique the dominant narrative. From this vantage, it became apparent that the privileges that I enjoyed were not God-given, natural, or even particularly well-deserved, but the fruits of a long history of extractive violence. It should be added that this violence has intensified to the extent that it is now difficult for the planet to supply the steady flow of cheap resources capitalism requires to keep expanding. In the effort to keep profitability up, capitalism has further eroded the already fragile welfare systems that kept the white middle class comfortable after the Second World War. Today, capitalism works extremely well for the one per cent of the world's richest, but not for many other people.

I also think that the fact that I grew up in a nation that can be described as semiperipheral to the core, or that at least exists at a significant political and geographical distance from the hegemonic core, was important.[1] I am sometimes asked how I can work on American culture while living in Sweden. I am certain that there are things that pass you by when you do not permanently inhabit the region that produces the culture you study, but such positionality also offers a certain critical distance. This book makes use of what the Warwick Research Collective (WReC, 2015) following Franco Moretti, has termed a new theory of world-literature. This is a type of literary study that argues that the same forces and

---

[1] Whether Sweden is best defined as a semiperiphery or as a part of the non-hegemonic core has been debated. Van Rossem (1996, p. 515) defines Sweden as semiperipheral while Arrighi and Drangel (1986) describe Sweden as part of the core.

the same inequalities that have shaped global society have also shaped literature and other types of culture. As discussed in more detail in the introduction to this book, this means that people at the centre of the core will have a notably different experience of what Immanuel Wallerstein calls the world-system than people in the semiperipheries, peripheries, or in the precarious margins of the core, and because of this, they will also tell very different stories. In other words, narratives produced in peripheral regions where crude oil, lithium, or diamonds are sourced will be very different from the stories told in the core where this oil is burned in passenger jets, where lithium exists in the batteries of Tesla sedans, or where diamonds are worn on engagement rings. By reading these different stories together, it becomes apparent that they are, in fact, telling stories about the same, combined, and unequal, world-system.

During the late twentieth century, Sweden enjoyed many of the privileges of the core and the driving of Teslas is far more common than the mining of diamonds within the borders of the nation. Indeed, many Swedish citizens have long enjoyed the privileges created by the nation's own 'internal' core–periphery relationship built around the extraction of iron, other minerals, lumber, and hydropower from land taken from the Indigenous Saami in the northern regions of Scandinavia. It was not difficult for a US culture praising whiteness and private property to find purchase. Even so, the US has always looked different when observed from Sweden and the rest of Scandinavia than when perceived from within its own borders. Socialism has never been anathema to the Swedish political conversation and, from the Scandinavian perspective, the US meddling in South American politics, and the bloody invasions of Vietnam, Granada, Iraq, and so forth, could easily be perceived as the ruthless bids for continuing world-systemic dominance that they were.

The theory of world-literature that the WReC has introduced builds on Marxist historiography and furthers a materialist understanding of how literature comes to narrate the world. Throughout the book, I also make use of scholarly work that can be labelled as eco-socialist, eco-feminist, postcolonial, decolonial, and Anarchist. It is important to recognize that there are significant tensions between these fields, as well as between scholars and movements within them, tensions that have at times taken the form of public disagreement. I understand and respect the endeavour to make certain positions clear and there is certainly a need for vigorous debate that hones our thinking. However, in this book, I have refrained from entering these polemics and from taking sides. This makes it possible to focus on the far more pronounced tension that exists between the different radical modes of thinking I make use of, and the militarized, neoliberal models that theorize capitalism and crisis as inevitable: between the attempt to prevent continued socio-ecological breakdown and the endeavour to save the capitalist system that has produced it.

The decision to turn to radical theory that centres ideology and material (colonial and capitalist) history, rather than to the posthuman or new materialist alternatives often used in ecocriticism, has to do with the fact that I believe that radical theory makes possible a much more direct engagement with the forces that are causing the biospheric crisis. If, as former Executive Secretary of the United Nations Framework Convention on Climate Change Christiana Figueres (2022) has argued, the 'climate crisis, the nature crisis, the inequality crisis, the food crisis all share the same deep root: extractivism based on extrinsic principles' (p. xviii), radical theory that centres extraction and that makes visible the capitalist system that depends on extraction, takes us straight to the problem. I do not think that any of the interrelated crises that Figueres mentions can be resolved if we do not centre extractive capitalism and if we do not consider our relationship to it. If literary scholarship refrains from participating in such centring, it is, at best, an interesting scholarly exercise, and, at worst, part of the effort to make capitalism and the violence it produces invisible. I realize that such claims may be provocative to (ecocritical) scholars whose work does not consider the history of capitalism and the impact it has had on the lives of peripheralized people and the nested socio-ecological world in some way, but my purpose with this book is ultimately not to provoke. It is to invite.

Kalmar and Växjö, Sweden   Johan Höglund

WORKS CITED

Warwick Research Collective (WReC). 2015. *Combined and Uneven Development: Towards a New Theory of World-literature.* Liverpool: Liverpool University Press.
Figueres, Christiana. 2022. 'Foreword'. In *Earth for All: A Survival Guide for Humanity*, edited by Dixson-Declève, S., Gaffney, O., Ghosh, J., Randers, J., Rockström, J. and Stocknes, P.E. Gabriola Island: New Society Publishers, pp. xvii–xviii.

Acknowledgements

This book would not exist had it not been for the RJ Sabbatical funding generously given by the Swedish foundation Riksbankens Jubileumsfond. This grant made it possible for me to compile a number of pre-existing ideas and pieces into a coherent manuscript. I am also very grateful to the Linnaeus University Centre for Concurrences in Colonial and Post-colonial Studies that co-funded my sabbatical, and to Magnus Bergvalls stiftelse and Stiftelsen San Michele who made crucial research trips and writing retreats possible. This book is Open Access thanks to funding provided by Riksbankens Jubileumsfond and by Linnaeus University's Open Access initiative.

I am furthermore enormously grateful to the many colleagues and friends who have generously commented on the manuscript along the way. I am particularly indebted to my brilliant co-worker Rebecca Duncan for her very incisive reading of much of the manuscript and for helping to steer the project in the right direction. I am similarly grateful to Rune Graulund who invited me to work alongside Anthropocene Aesthetics at the University of Southern Denmark during a portion of my sabbatical, who made sure I never wanted for friendship when in Denmark, who commented on drafts and who kept my phone buzzing with supportive texts when I needed them the most. Many thanks also go to Niklas Salmose for many excellent conversations about climate narratives, for comments on the project in its various phases, to Mike Classon Frangos for generously reading a chapter despite being on parental leave and to

Emily Hanscam for useful feedback on an early draft of Chapter 1. This book is also better than it otherwise would have been thanks to comments from Karl Emil Rosenbæk at the University of Southern Denmark, Lucia Hodgson, guest research fellow at Linnaeus University and Isabelle Hesse at the University of Sidney. Finally, my thoughts go to my friend Justin D. Edwards who helped steer this project in the right direction during its early stages, but who departed this world before it was completed. He is sorely missed.

I have also benefitted greatly from being able to present material before several research groups, most importantly the Aesthetics of Empire research cluster that is part of LNUC Concurrences, and the research seminar for film and literature at Linnaeus University. I also want to thank Åsa Bharati Larsson who invited me to present at the Swedish Institute for North American Studies (SINAS), Zlatan Filipovic who invited me to present part of the project at the Gothenburg University research seminar, and, again, Rune Graulund who let me talk to participants at the Danish 2023 PhD school 'Environments: Extinct, Envisioned, Evolving' at Sandbjerg. At all of these seminars, I received generous and constructive feedback on various parts of the project and the book would be very different without it.

In addition to this, I need to thank my editor at Palgrave Macmillan, Molly Beck, and the editors of the New Comparisons in World Literature book series Pablo Mukherjee and Sharae Deckard, for giving the book their support and the best possible home. Thanks also to the anonymous reviewers who have provided feedback on the manuscript at various stages and to Lobke Minter for all her hard work (at all days of the week) on editing the sometimes unruly language of this book. I am also grateful to Stefan Amirell for helpful feedback on the RJ Sabbatical application that made it possible for me to devote time to this project.

None of the chapters included in this book have been published before, but the first chapter of this book has benefited from the work I did, and the editorial comments I received, on the chapter 'Settlement and Imperialism' published in the *Cambridge Companion to American Horror* (Cambridge UP, 2022), edited by Stephen Shapiro and Mark Storey, and on the article 'Challenging Ecoprecarity in Paolo Bacigalupi's Ship Breaker Trilogy' that was part of a special issue called Challenging Precarity published by *Journal of Postcolonial Writing*, edited by Janet M. Wilson, Om Prakash Dwivedi and Cristina M. Gámez-Fernández. My thinking on the topic of the book has also been aided by the editorial commentary I received as I wrote the chapters 'Globalgothic, Viruses and Pandemics' published in *The Edinburgh Companion to Globalgothic* (Edinburgh UP, 2023) edited by Rebecca Duncan, and '*Maggie* in the Necrocene' published in *The Anthropocene and the Undead* (Lexington Books, 2022) edited by Simon Bacon.

In addition to this, I must thank the people who have kept me happy and who have put up with me, or with my absence, during especially intense writing periods. Thanks to Nanna Nerest for her very rewarding and lasting horror friendship during long research visits in Copenhagen. Thanks also to Om Dwivedi, Agnieszka Soltysik Monnet, and Liselotte Åström for support, encouragement, and important conversations along the way. Finally, my thoughts and gratitude go to David, Agnes, Edith, Hilda, and to my mother and father, who have put up with a constantly writing father and son. Most importantly, thanks to my wife and life companion Cissi, for your never-wavering support, friendship, patience, and love. This book is just one of many things I could not have done without you.

CONTENTS

LIST OF FIGURES

CHAPTER 1

Introduction: The American Climate Emergency Narrative

In *Annihilation* (2014a), the first novel of Jeff VanderMeer's Southern Reach Trilogy, a territory known as Area X has been sealed off by a secret, military organization called Central. They have set up a military research station, the eponymous Southern Reach, and from this vantage, they are trying to both contain and explore an increasingly unpredictable and dangerous territory. Plants and animals behave in strange ways, the landscape is inconstant and impossible to map, and the humans who enter it soon begin to exhibit weird and erratic behaviour. *Annihilation* follows a biologist who is part of one of these expeditions. She soon finds herself under threat both from entities within Area X and from other members of the expedition, even as she is turned into something probably not quite human. *Authority* (2014b), the second instalment to the trilogy, tells the story of John 'Control' Rodriguez, a newly appointed director of the Southern Reach facility. He is from a family that has, for a long time, helped run the clandestine operations of Central, but his attempts to manage the station, and to understand and contain Area X, are fraught with failure. Area X resists all military and scientific strategies, it is expanding beyond the geographical, ecological, human, and civilizational borders guarded by the station. In this future, the natural world is out of control.

Miranda Iossifidis and Lisa Garforth (2022) are two of many scholars who have read *Annihilation* as 'climate fiction', because 'the uncanny

© The Author(s) 2024
J. Höglund, *The American Climate Emergency Narrative*,
New Comparisons in World Literature,
https://doi.org/10.1007/978-3-031-60645-8_1

atmospheres brought to life by the text, and the affective responses of some of its readers, create new ways of imagining climate futures' (p. 248). According to standard ecocritical scholarship, climate fiction is a new literary genre primarily involved in the description of the dark futures that will ensue if nothing is done about climate change. As argued by Imogen Malpas (2021), 'climate fiction provides us with the beginnings of the roadmap we so sorely need to achieve a global society that is both abundant and sustainable' (np). Climate fiction is supposed to be able to provide such roadmaps partly because it is based on existing climate science. In the words of Gregers Andersen (2020), climate fiction depicts 'worlds resembling those forecast by the IPCC' (p. 1), and, by doing so, they provide 'speculative insights into how it might be to feel and understand in such worlds' (p. 1). Borrowing a concept launched by geologists (who get to name climate events and epochs), climate fiction scholars sometimes also refer to the genre as Anthropocene fiction (Trexler 2015). The Anthropocene concept here identifies the human as the entity that has produced the event typically referred to as the climate crisis. Thus, climate fiction is typically read and studied as a genre that takes people into futures predicted by climate science, where human activity has caused extensive flooding, parching drought, and terrible storms, and where conditions for human life have been drastically altered.

Before considering how the Southern Reach Trilogy corresponds to these requirements, it is necessary to contemplate the widely circulated, and increasingly problematized, claim that the ongoing breakdown is caused by the human as a species. Since atmospheric chemist and Nobel Prize winner Paul Crutzen and biologist Eugene Stoermer (2000) launched the concept of the Anthropocene, it has come under fire from those who deny the very notion that the planet is warming in dangerous ways, but also from a growing number of scientists, sociologists, environmental historians, Indigenous scholars, and climate activists who find the concept, and the history of biospheric breakdown it relies on, misleading. While this contingent strongly agrees that human *activity* has caused a biospheric crisis, their crucial objection is that this crisis has not been produced by all humans equally, but rather by a particular group of people in the Global North who have long taken advantage of the opportunities offered by colonialism and capitalism. These critical voices include scholars such as John Bellamy Foster (1999), Jason W. Moore (2015, 2016), Andreas Malm (2016), Malm and Hornborg (2014), Hannah Holleman (2018), Macarena Gómez-Barris (2017), Katheryn

Yusoff (2018), Simon Lewis and Mark Maslin (2015), and Heather Davis and Zoe Todd (2017).[1] Work by these authors employ different terminology and perspectives and disagree on certain points, but they share the understanding that the history of the climate crisis is also the history of extractive (colonial) capitalism. In this way, climate breakdown has its origin in the extractive monocrop plantations of colonized America, in the elimination of Indigenous people that made the plantation possible, in the chattel slavery system that serviced the plantation, in the industrial landscape created to refine, transform and market the proceeds of the plantation, in the subterranean worlds where the coal and oil necessary to power industry are sourced, and in the establishment of vast armies tasked with protecting this development. Some of those who make this argument still employ the term Anthropocene, but many have turned to the alternative concept of the Capitalocene, coined by Andreas Malm but theorized primarily by environmental historian Jason W. Moore. This book draws from several of the thinkers referenced above, but it is primarily informed by Moore's work and understanding of the past, present, and future of the ongoing crisis.

I will return to this crucial rethinking of what drives a breakdown that is both social and ecological, rather than just climate-related, in this introduction and in the chapters that follow. First, though, I want to reconsider the idea that the Southern Reach trilogy is 'climate fiction' and that, as such, it is capable of showing us the road towards more sustainable ways of life. While there is certainly some merit to this observation—the novel may indeed, through its representation of nature as weird and disturbing, problematize the prevalent notion that humans are in control of, or apart from, nature—it must also be noted that this novel (like many other so-called climate fictions) is strangely quiet on what has caused the weird ecological and human changes that take place within its pages. In the Southern Reach trilogy, the entity that has initiated the transformation of ecology in Area X is not capitalism, fossil-fuelled human industry, or even humanity generalized into species, but an undefined, occult, or possibly

---

[1] The claim that 'rentier capitalism' fuels the climate crisis has also been made in climate science publications such as Dixson-Decléve et al.'s *Earth for All* (2022). In addition to this, the 2022 IPCC report (IPCC 2022) recognizes the role that industrial capitalism and colonialism have played in the climate crisis, especially in the Global South. See, for instance, section 4.3.8 'Observed Impacts on the Cultural Water Uses of Indigenous Peoples, Local Communities and Traditional Peoples' (pp. 595–596).

extra-terrestrial entity. Against such an otherworldly agent, the characters and institutions of the novel can do little. It must also be noted that the trilogy does not register how the biospheric breakdown it references is part of a global, interrelated, social, economic, and ecological metacrisis, and it does not acknowledge how unevenly biospheric breakdown affects different communities across the planet; how for some, it is an anxiety-inducing future and for others an already brutal lived experience (Nixon 2011; Sultana 2022). This lack of history and global context adds to the confusion experienced by the characters of the trilogy. They cannot comprehend what is going on because the novel refuses to make the world it imagines comprehensible. At the end of the day, the only form of action the characters can take, as the title of the final novel *Acceptance* (2014c) suggests, is to acknowledge that ecology will be fundamentally different from now on, and to figure out how to live in this drastically altered world.

In other words, when read as climate fiction, the Southern Reach trilogy makes what ecocritic Erin James (2015) has called a 'storyworld' in which biospheric breakdown is allegorized into an uncanny spectacle, where the systems, people, and histories that are responsible for this breakdown go unrecognized and where the unevenness of the slow and fast violence of climate change is elided. What geologists have named the Anthropocene may loom in between the lines, but *Annihilation* does not evoke a future that resembles the predictions of the IPCC, and hardly provides readers with a roadmap capable of taking them to an abundant and sustainable global future. If climate fiction is supposed to do this, *Annihilation* does not seem to belong to the genre. It is oblivious to the history that has produced socio-ecological breakdown, and it is neither about the biospheric futures predicted by climate science nor about the uneven experience of planetary-scale socio-ecological breakdown in the present.

This is fine, of course. This book does not ask that stories such as *Annihilation* are not written or appreciated. The Southern Reach Trilogy is a remarkable literary achievement and my purpose is not to create borders for what storyworlds fiction should be allowed to make. That noted, to call the Southern Reach Trilogy a story about biospheric breakdown is problematic because its depiction of such breakdown is ahistorical, unscientific, and ignores how unevenly it affects different people. However, the trilogy's evocation of anxieties tied to eroding social and ecological worlds, and its portrayal of how the militarized security state responds to such uncanny erosion, are precise and jarring. The Southern Reach trilogy may not be saying anything about why the biosphere is eroding, it may not help readers understand how to mitigate such erosion, and it is notably silent on the fact that this breakdown is experienced in radically different ways in different parts of the world, but it is certainly vocal on how the US security state considers biospheric erosion the beginning of a potentially *terminal crisis* that merits the declaration of an *emergency* and a state of exception.

This book shows that this is a common strategy in much American fiction set in futures transformed by socio-ecological breakdown. While there are important novels and films that acknowledge the role that capitalism has played in such breakdown and that recognize how unevenly breakdown is experienced (as discussed in the final chapter of this book), there is a notably large body of texts that instead centres on capitalist crisis, emergency, and the securitization strategies employed to address them. VanderMeers' trilogy approaches this type of text by describing how a seemingly sudden biospheric crisis creates a considerable dearth of security for a US national security apparatus charged with maintaining the conditions under which US capitalist society can work smoothly. Similarly, it is about how this disturbance prompts the establishment of *a state of exception* that provides the organization in charge (Central) with far-reaching and unconstitutional powers. In this way, the Southern Reach trilogy is most usefully analysed as a narrative that registers how socio-ecological breakdown is an emergency for the nation-state, and for the security institution tasked with protecting this state and its citizens. In view of this overarching concern, the Southern Reach Trilogy is better described as an *American Climate Emergency Narrative*, than as climate fiction.

This may seem like a minor adjustment in terminology, but it does transform the scholarly engagement with this type of text. To name texts such as the Southern Reach trilogy *American* Climate Emergency Narratives recognizes that these texts come out of, and communicate with, a particular location within what Immanuel Wallerstein (2004) has called the capitalist world-system. In other words, these narratives do not enter the conversation on socio-ecological breakdown out of a universal nowhere. With the help of the world-literature perspective introduced by Franco Moretti (1996, 2000) and modified by the WReC (2015), these narratives can be perceived as texts written or otherwise produced within and for the core defined as a privileged and dominant sector of the world-system. This means, as I discuss in more detail below, that they speak from, and about, a certain privileged experience of this system. To call them American *Climate Emergency* Narratives is furthermore to centre how these texts narrate the security strategies that are used in the US and in other parts of the world to manage and adapt to, rather than mitigate, the various crises and disasters that biospheric erosion is accelerating.

With the help of the concept of the American Climate Emergency Narrative, this book shows how socio-ecological breakdown is often described as an existential crisis for American capitalist modernity, rather than for the planet. As the first chapters of the book demonstrate, this is a narrative that emerges out of the same history as socio-ecological breakdown itself, and that has evolved alongside the economic, ecological, military, and labour-related crises that accompany capitalism. In its most recent, intermedial form discussed in the second half of the book, it registers, but also contributes to, capitalism's attempt to secure its own future at a time of epochal capitalist crisis. The book furthermore shows that while some of the texts I describe as American Climate Emergency Narratives signal an awareness of the fact that global warming and biospheric erosion are the result of extractive capitalism, they struggle to imagine any other social order and typically turn to capitalism, and to security mechanisms such as the US Department of Defense (DoD), as the only entities capable of relieving their plight. In this narrative, militarized capitalist modernity may have contributed to the emergency at hand, but it is also somehow the solution. In this way, much of the fiction that is discussed in this book sanitizes the idea that capitalist nations must respond to what is imagined as an anthropogenic climate emergency through military adaptation; through the creation of organizations such as Central and the establishment of places like the Southern Reach, via participating in local and global resource wars, by containing migrants displaced by climate change, even by making war on eroding and uncooperative ecology itself. Thus, the American Climate Emergency Narrative I study is not primarily an attempt to give affective shape to recent IPCC reports, but instead a part of both the capitalist and colonial history that has produced the present crisis, and of the culture that attempts to manage this crisis.

WORLD-SYSTEM, WORLD-ECOLOGY, CHEAP NATURE, AND CRISIS

To come to terms with the American Climate Emergency Narrative—to understand where it comes from, what it is that it registers, obscures, celebrates, and condemns, and how it envisions events such as crisis, emergencies, and responses such as securitization and militarization—it is necessary to consider the claim that capitalism is the engine of socio-ecological breakdown in more detail and to explore the material history that bears this claim out. It is similarly important to probe the instrumentalizing concepts of 'climate crisis' and 'climate emergency' and try to understand how they function when contrasted to the very different understanding of crisis and emergency that radical eco-socialist, decolonial, and Indigenous methodologies encourage. In what follows, I briefly describe the historical and intellectual framework that guides my analysis of the origins, rhetorical evolution, and central concerns of the American Climate Emergency Narrative. I also discuss the capitalist imperative to secure the extractive enterprise on a local and global level, and the role that emergency has played for such securitization.

An important beginning to grasp the role that capitalism has played in biospheric breakdown is Wallerstein's aforementioned concept of the capitalist world-system. This concept notes how European nations reshaped the world into peripheries from where resources were sourced cheaply, a core where these resources were commodified and their proceeds accumulated, and semiperipheries where both processes occurred. The formation of this world-system was complex and involved wars, trade agreements, the formation of new nations states, and so on, but one crucial phase in the making of the world-system that merits particular attention here was the privatization and often violent enclosure of land in Europe and then abroad. As James Alfred Yelling (1977), Jeanette M. Neeson (1993), and Robert P. Marzec (2002) discuss, the enclosure movement began in Britain in the late medieval period and transformed the predominately rural population of Europe into landowners or labourers employed (or forced) to work the land they previously held in common with other

people. As Marzec observes, these enclosures were 'England's prepara-
tory experiments in the colonization of its own land before the nation
began to acquire land abroad' (2002, p. 13).[2]

The establishment of enclosures abroad, and thus of the commodity
peripheries that make up a vital segment of the world-system, was greatly
energized by the 'discovery' of America in the fifteenth century. As I
discuss in more detail in Chapter 2, this 'New World' presented early
European capitalism with vast tracts of enormously fertile land to enclose
and reorganize not simply into settlements, but into plantations that grew
cash crops for a European market. The establishment of such enterprises
was made easier by the fact that the first settlers brought illnesses for
which the Indigenous population had no immunity. By the beginning
of the seventeenth century, an estimated 90 per cent of the population
had succumbed to these illnesses (Koch et al., 2019). This left the continent relatively open and enabled the comparatively speedy establishment
of the extractive, triangular economy that brought cash crops grown in
the New World to Europe, trade items to Africa, and enslaved people
to the plantations in the New World. In the centuries that followed, the
adaptability of capitalism to various crises (many of its own making), its
ruthless transformation of people and land into commodifiable resources,
turned capitalism into the only game in town: into the world-system.

Wallerstein's detailed history of the origin and development of the capitalist world-system has influenced more recent world-system authors such
as Giovanni Arrighi (1994), decolonial scholars such as Aníbal Quijano
(see Quijano 2000; Quijano and Wallerstein 2000) and environmental
historians such as Jason W. Moore. Quijano is one of several decolonial
thinkers who has done work on how colonial capitalism transformed living
conditions and social realities in South America and other parts of the
Global South, and on how this reshaping is still going on. Building on
but also modifying Wallerstein's and Arrighi's work, Moore has crucially
argued that capitalism did not simply reshape economic and social systems
across the world, it also remade the *ecology* of the planet to suit its

---

[2] This is why Wallerstein and Moore, and other Marxist, decolonial, and postcolonial
scholars such as Bellamy Foster (2002), Anibal Quijano (2000), Benita Parry (2004),
and Neil Lazarus (2011), have argued that capitalism and the colonial project have been
interdependent to the extent that they need to be thought of as the same thing. Thus,
Donald Denoon (1983) talks about 'settler capitalism', Anibal Quijano (2000) discusses
'colonial capitalism', and, in a recent essay, John Bellamy Foster, Brett Clark, and Hannah
Holleman (2021) write 'capitalism/colonialism'.

needs. Moore thus names the profoundly interrelated ecological, social, and economic world that capitalism has made over the past 400 years the *capitalist world-ecology*. An important point made by Moore is that capitalism has not simply worked *on* nature but has always been *in nature* so that by reorganizing ecology it has also reshaped its own conditions and possible futures. As I will return to, this is a crucial observation that has great bearing on the attempt to understand how capitalism has helped to shape the current socio-ecological moment and thus also what I term the American Climate Emergency Narrative.

Since the inception of the capitalist world-ecology, capitalism's reorganization of the social and ecological world has revolved around the appropriation of what Moore calls commodity frontiers (Moore 2015, p. 63) where the cost of growing crops, mining metals, and minerals, or extracting fossil energies, has been relatively low. As Patel and Moore (2018) argue, 'capitalism has thrived not because it has been violent and destructive (it is) but because it is productive in a particular way' (p. 19). This particularity is intimately tied to its ability to adapt to crises and to thus continue to source nature and labour *cheaply*. To do so, Moore argues, has been the guiding imperative or 'law of value' (Moore 2015, p. 24) of capitalism. As an example, acting according to this law of value, actors within the early capitalist world-ecology preferred to relocate the sugar plantation from a depleted Madeira to Brazil and the Caribbean, where the necessary fuel and fertile soil could be found, rather than restoring the ecology of Madeira. It was not only non-human nature that was made cheap in this way. To capitalism, 'virtually all peoples of color, most women, and most people with white skin living in semicolonial regions' (Moore 2016, p. 91) also belonged to the realm of nature. As such they were also exhausted and replaced when convenient. In this way, capitalism has progressed by turning the planet into a source of what Moore terms *Cheap Nature*.

As can be seen from the Madeira example, treating people and nature cheaply does not leave room for recuperation and restoration. The law of Cheap Nature depletes people, societies, land, seas, mountains, and soils, and in doing so, it also changes the conditions for the global, capitalist project. In the past, capitalism has been able to manage the depletion of nature and labour by relocating its extractive ventures to other sites, and by reorganizing its relationship to labour. Indeed, as Arrighi (1978, 1994) has shown, and as Naomi Klein (2007) has also influentially argued, crises are endemic to capitalism. Such crises include mines becoming depleted, banks collapsing, commodities going out of fashion, and (Indigenous) communities resisting the transformation of rainforest to grassland for cattle or palm oil plantations. However, while a capitalist entity (a corporation, group of owners, a bank) may suffer from a particular crisis, the capitalist system not only produces crises, it also exploits them. Unemployment makes labour cheaper, resistance from Indigenous groups or competing nation-states boosts the sale of military hardware, and deregulation designed to stimulate energy production opens new areas up for extraction. On Madeira, the end of the sugar plantation became the beginning of the wine industry (Patel and Moore 2018, pp. 17–18).

However, the world has now reached a point where such deregulation, relocation, and reorganization are becoming very demanding or even impossible. The deeply interrelated, uneven, and fundamentally extractive social, economic, and ecological relations on which capitalism relies have reduced the entire planet's capacity to recuperate. As Moore (2015) puts it 'it is increasingly difficult to get nature—including human nature— to yield its "free gifts" on the cheap' (p. 13). At this moment in time, agents of the capitalist world-ecology are trying to adapt to and manage this elevated crisis, but the global capitalist antidote to deregulate and capitalize—to tweak the neoliberal model so that labour is more accessible and cheaper (the gig economy), to (re)introduce new techniques that enable more effective extraction (fracking), and to open new territories for exploitation (the Arctic, new sections of the North Sea, rainforest in South East Asia and South America)—are clearly short-term solutions. To Moore (2015), this indicates that 'we may be experiencing not merely a transition from one phase of capitalism to another, but something more epochal: the breakdown of the strategies and relations that have sustained capital accumulation over the past five centuries' (p. 13). Should these strategies and relations truly break down, this would spell the end of the law of Cheap Nature and of the capitalist world-ecology.

This suggests that there is a need to consider two intimately entangled but actually very different, and differently lived, crises. One is planetary, ecological, and biospheric, and the other systemic and capitalist. The planetary, biospheric crisis is considerable; a mass extinction of species at a rate not experienced since the demise of the dinosaurs 64 million years ago (see Kolbert 2014; Ceballos et al. 2015). The capitalist crisis is similarly vast but while it is intimately connected to the biospheric crisis, *it is not the same thing*. The epochal capitalist crisis is not triggered by the fact that the biosphere is eroding as much as it is produced by the increasing lack of new Nature to appropriate and work *cheaply*. For the capitalist world-system, *this shortage is the crisis*. In the postcolonies of the Global South, the crises produced by extractive capitalism have been felt for hundreds of years. To the affluent, most of which are ensconced in the Global North, it appears as something new and sudden: As a *future crisis* rather than the culmination of a long and extractive history.

This account of crisis and emergency is central to the study of what I call the American Climate Emergency Narrative. Novelists, filmmakers, and other cultural workers and entrepreneurs who produce this kind of text may project the sense that the stories they tell are about the biospheric crisis, when they are in fact exploring the emergencies that erupt out of the epochal capitalist crisis as this takes shape in the Global North. As I have suggested, and as the coming chapters reveal, the American Climate Emergency Narrative is fundamentally uninterested in, even oblivious to, the capitalist history that has brought the current socio-ecological crisis on, and in the fundamentally uneven experience of this crisis in the present. It is instead obsessed with the possibility that capitalism might end, with the various threats to this world order that biospheric breakdown may produce, and with the need to securitize capitalist modernity. Thus, the American Climate Emergency narrative may appear as a textual form that calls attention to biospheric breakdown, and ecocritical scholarship may take it at its word, but as this book argues, the crisis that it probes is fundamentally capitalist.

EMERGENCY AND SECURITIZATION

The possibility that the capitalist world-ecology may face an epochal crisis is central to the leveraging of the emergency concept in the present, and to a host of security mechanisms designed to prevent or postpone the crisis. As the concept of the American Climate Emergency Narrative implies, this is a body of texts that centres emergency, probes the various insecurities that constitute emergency, and proposes solutions that will resolve the emergency and the crisis that has provoked it. As described in the previous section, the crisis that merits the declaration of the emergency and the action taken to relieve it is not biospheric breakdown per se, but the possibility that the conditions that keep the capitalist world-system running will cease to be.

This has been a concern for core actors within the capitalist world-ecology since its early beginnings. Importantly, the creation of enclosures in the old and new world was also the creation of a new type of insecurity, and thus the beginning of a militarized consciousness that Robert P. Marzec (2015), adapting Michel Foucault's (1991) notion of 'governmentality', has termed 'environmentality' (p. 4).[3] The enclosure effectively transformed what used to be a sustainable, multispecies, communal habitat into a specific energy or commodity frontier that a landowner or other polity now depended upon for accumulable income. This, as Marzec observes, 'introduced modern conceptions of privatization, surveillance, and environmental manipulation' (p. 11). What needed to be kept secure, from this moment on, was not humans (however defined) or animals or land, but the enclosure and the commodity frontier that was contained within this enclosure. It can thus be argued that what was securitized by the private actors and the governments that made up the early colonial state was extraction (of sugar, grain, livestock, gold, coal and oil, labour, i.e., of humans and extra-human nature) itself. Put differently, the primary responsibility of state and private militias in America became the securitization of the law of Cheap Nature.[4]

The securitization of the commodity frontier had enormous consequences for how human and non-human lives were lived and ended in early America. From the establishment of the first enclosures and colonies, biophysical nature, and many (racialized, Indigenous) people (Moore 2016, p. 79) became something to be governed and managed. At the same time, the enclosure and privatization of land made it possible to work both the land and the people who tended it harder; to force land and labour to yield more than they had previously done and to expand the extractive process to new regions. To conquer and enclose additional land and to govern and manage this land—to reorganize nature into a capitalist world-ecology—it is necessary to exert a great deal of violence. There must be a mechanism that can manage the various emergencies that threaten extraction, that keeps property borders intact, guarantees the extraction of value from the land, eliminates Indigenous people,

[3] Marzec thus extends the notion of 'governmentality' to describe how the modern state extends control not simply over the human body, but also over the environment. Just like governmentality transforms/secures the human body into an entity that can be (self) disciplined and effectively conditioned by the state, environmentality produces nature as a repository of energies vital to the state.

[4] Or, to be more precise, what Moore terms the 'Four Cheaps': 'labor-power, food, energy, and raw materials' (Moore 2015, p. 27).

controls rebellious labour in fields and industries, and defends, or steals, already enclosed land from competing nation-states. This mechanism, as Rosa Luxemburg (2015) and Louis Althusser (2014) have argued in different ways, and as Alex Vitale (2017) and William I. Robinson (2020) have recently discussed, is the military and the police. By expanding and securing the borders that mark enclosures, and by keeping labour within these borders under control, the military—in its private paramilitary form, or as funded by local governments or by states or nation-states—performs an essential service for capitalism.

This is important when considering the present epochal crisis and the leveraging of emergencies. It is similarly important when analysing the storyworlds that are produced within the American Climate Emergency Narrative. The tendency to turn to the armed sections of the security apparatus when emergencies arise should cause concerns at a time when a number of cities, nation-states, and supranational organizations have declared 'climate emergencies'.[5] Because such declarations oppose climate change denial, they are often welcomed by climate scientists and activists. Indeed, there is every reason to consider the current erosion of the biosphere and the ongoing sixth mass extinction of species as precisely an emergency for the planet and the many species that inhabit it. However, declarations of emergency do not necessarily lead to attempts to transform the root causes that have created the emergency. Rather, as Mike Hulme (2019) has observed, emergencies are often a reason for states and other actors to establish what Giorgio Agamben (2005) has called *states of exception* designed to manage the 'conditions of war, insurrection, or terrorist threat' (p. 23), but not to alter the system that has produced these conditions. In fact, as Rebecca Duncan et al. argue (2023), 'by framing climate change as an imminent but manageable disaster, it becomes possible to bypass questions relating to where this disaster comes from, thereby licensing solutions that reproduce, and work in the interest of, the extractive (colonial) system' (p. 477). In other words, there is a considerable risk that declarations of climate emergency introduce policies that centre the needs of the states that have declared them, and of the capitalist world-system, rather than the needs of the planet and its many people. To declare (climate) emergencies that ultimately serve to strengthen the system that has produced them not only decentres the entities that are

---

[5] The first place to declare a climate emergency was City of Darebin, Melbourne, Australia in 2016. The European Parliament followed suit in 2019.

experiencing crisis more keenly (the planet and precarious communities in various parts of the world), it risks *accelerating* both the biospheric crisis and already existing inequalities.

This is borne out by the fact that the regions, nations, and unions that have so far declared formal climate emergencies have, as a rule, been the least affected by ongoing socio-ecological breakdown and are the least likely to suffer the consequences of accelerating biospheric erosion (see Birkmann et al. [2022] for an estimation of the relative vulnerability of different nations to socio-ecological breakdown and Cedamia [2022] for a list of the nations and regions that have declared climate emergencies). By contrast, very few of the nations that are estimated to be especially vulnerable to climate change have declared climate emergencies. The few nations in the Global South that have made such declarations are mostly islands that have already begun to disappear into rising oceans. There is a clear logic to this discrepancy in declaring climate emergencies. For many people in the Global South, living within nation-states established by, and still mostly dependent on, the nations that make up the core of the world-system, the crisis began hundreds of years ago. What would be the point of declaring an emergency now? In addition to this, from the perspective of the core of the world-ecology, for nations in the Global South to do so risks compounding the emergency, since such a declaration may curb access to the commodity frontiers located in this region. In any case, forcibly inserted into the capitalist world-ecology as commodity frontiers where nature and labour are sourced on the cheap, 'developing' nations in the Global South simply cannot afford to declare climate emergencies.

In the Global North, and also in nation-states that aspire to supplant the current hegemonic core of the world-system, a different story is playing out. Whether these nations have declared emergencies or not, they are mobilizing their security apparatuses to adapt and manage oncoming crises. As described in considerable detail by Marzec (2015), dominant nation-states and their militarized security apparatuses are managing the biospheric crisis by projecting climate change as an 'engagement opportunity' (p. 9). Indeed, as Marzec has also observed, the biospheric crisis is perceived by the US State, and by institutions such as the US DoD as a 'threat multiplier' (p. 2) and as such it is understood as an emergency *for the US nation-state*, and for the global capitalist economy this state relies on and directs, rather than for the deeply interconnected multi-species ecology that keeps the planet healthy, or for the millions of people who live precarious lives in the shadow of accumulation.

The tendency to seek to resolve capitalist crisis by leveraging emergency, and by preparing to do violence in order to resolve this emergency, is crucially important to the study of the American Climate Emergency Narrative. Writing from the core that narrates capitalist crisis via images of the escalation of internal and international tensions into massive migration movements and warfare is haunted by this militarized history. When such writing turns to the military as a universal panacea for the insecurities that give direction to the narrative, (military) adaptation is perceived as the only viable strategy forward. No other future history than the continuation of the present system can be imagined under these conditions. Some of the texts discussed in this book may correctly identify militarized capitalist modernity as the engine of climate breakdown (and of related forms of social erosion), but even then, alternatives to this social and military order are perceived as impossible or as so detrimental that oblivion is preferable. Thus, the American Climate Emergency Narrative is ultimately about the need to restore and repair capitalist modernity, even if such restoration (absurdly) involves making war on the planet itself.

CULTURAL STUDIES IN THE CAPITALOCENE: THE WORLD-LITERATURE PERSPECTIVE

Just as the realization that fossil-energized capitalism is driving biospheric breakdown 'changes everything' (Klein 2015) for how the climate crisis is understood and acted upon, it also transforms the conditions for cultural theory. As Andreas Malm proposes in *The Progress of this Storm* (2018) 'any theory for the warming condition should have the struggle to stabilise climate – with the demolition of the fossil economy the necessary first step – as its practical, if only ideal, point of reference' (p. 18). While there is certainly ecocritical work that openly centres the need to dismantle capitalism, there is a notable tendency in ecocriticism (much of it produced in the US and Europe) to flatten the global impact of socioecological breakdown, to erase the material history that has produced it and to take for granted that the (future) European or US experience of this breakdown is somehow universal. A large number of literary ecocritical studies also focus on the *affect* produced by the reading of climate narratives, rather than the *effects* that capitalism has on human relations and ecology, on the combined and uneven world-ecology, and on the climate narrative as such.[6] A better starting point for ecocritical theory at a time of socio-ecological crisis is arguably to acknowledge first that most climate narratives are produced from within, and also for, a particular location of the world-system or world-ecology, and then that these narratives are read from very different positions. A point worth repeating is that the authors, directors, and production companies of the American Climate Emergency Narrative are, for the most part, and not surprisingly, located in the US. This means that they emerge out of a particular, core position of the world-system, and also out of the specific ecological, material, and narrative history that has been central to the production of the current socio-ecological crisis.

By reading the American Climate Emergency Narrative as a culture that emerges out of a particular location of the world-system, this book builds on a fairly recent development in world-literature studies introduced by Italian literary historian Franco Moretti (1996, 2000) and further developed by the members of the Warwick Research Collective (WReC). Building on Immanuel Wallerstein's world-system model, and recalling Fredric Jameson's point in *The Political Unconscious* (1981) that narrative is a 'socially symbolic act' (p. 1), Moretti observes that capitalism has created very different conditions for life in the core, the peripheries, and semiperipheries of the world-system, and that these conditions inform literature in a fundamental way. Extending Moretti's work, the Warwick Research Collective has argued in *Combined and Uneven Development: Towards a New Theory of World-Literature* that 'the dialectics of

---

[6] When Alexa Weik Von Mossner (2017) suggests in her article 'Sensing the Heat: Weather, Water, and Vulnerabilities in Paolo Bacigalupi's *The Water Knife*' that climate fiction is a genre that allows 'readers to imaginatively experience *what it is like* to live in a climate-changed world' (p. 174, italics in the original), this statement assumes that readers inhabit a world where socio-ecological breakdown is still in the future. In this way, the US' (or the Global North's) experience of (future) climate change becomes universalised as everyone's experience. Furthermore, Mossner may be perfectly correct in arguing that 'climate fiction' that uses 'anthropogenic climate change as a catalyst for drastic developments in the ecological, economic and social realm [...] invites readers to understand on a *visceral* level that changed climatic conditions will inevitably lead to [...] conflicts and vulnerabilities' (175, italics in the original). However, if what scholars such as Mossner study is not really fiction about the (planetary) climate, but texts that register emergency and epochal capitalist crisis from a particular location of the world-system, this fiction, and critical interventions such as Mossner's, may not be the most helpful in the attempt to forge intellectual positions from where the work of dismantling eco-erosive capitalism can begin.

core and periphery [...] underpin all cultural production in the modern era' (2015, p. 51), and that 'world literature' should be understood as the 'literature of the modern capitalist world-system' (p. 8). To differentiate literary studies that focus on the world-system from other types of world-literature studies, the WReC suggest that 'we should begin to speak of "world-literature" with a hyphen' (p. 8). This book adopts this understanding of (and this way of spelling) world-literature.

The WReC and scholars (many belonging to this collective) such as Michael Niblett (2012, 2020), Sharae Deckard (2019), Stephen Shapiro and Philip Barnard (2017), Treasa De Loughry (2020), Jennifer Wenzel (2019), and Rebecca Duncan (2022) have also expanded Moretti's word-literature model by introducing the world-ecological perspective into it. Thus, Niblett describes in *World Literature and Ecology: The Aesthetics of Commodity Frontiers, 1890–1950* (2020) how world-literature does not respond just to the social and economic inequalities that capitalism produces, but also to the related, large-scale ecological upheaval that capitalism's pursuit of Cheap Nature causes. Niblett demonstrates how this happens through a study of literature produced at four different commodity frontiers (sugar, cacao, coal, and oil) that are located at different times and geographical locations, but that all serve an expanding capitalist world-system/world-ecology. His study shows that texts from these dispersed frontiers register extractive human and ecological effects in very similar ways. Descriptions of 'the sugar mill remorselessly consuming cane, sucking out the energies of its workers' are thus 'echoed' in narratives about the coal pit, the cocoa refinery, and the oil derrick (p. 2). In essence, literature from the periphery or semiperiphery registers the experience of being caught up in the extractive machinery of the capitalist world-ecology.

While sociological world-system research has tended to centre the core, much of the work by world-literature scholars such as the WReC, Niblett, Duncan, Sánchez Prado (2019), and others is comparative and focuses on the peripheries and semiperipheries of the world-system. Like the postcolonial literary studies that it complements and converses with, much of this scholarship thus sheds light on how cultural production in colonies and postcolonies have registered and responded to the (racialized and gendered) socio-ecological violence conducted in these sections of the word-system. This book adopts this theoretical perspective but extends it in a different direction. Instead of focusing on how the capitalist world-system is registered in the extractive peripheries or semiperipheries, it explores literature, films, and games produced within the *core* of the world-system.

To do so is important for many reasons. As described, socio-ecological breakdown is a consequence of the capitalist imperatives directed from the core, and it is also where the effort to prevent the breakdown from unfolding must begin. In addition to this, as Andersen rightly notes, North America is the 'main producer of climate fiction' (2020, p. 7). By focusing on this geographical, economic, and cultural section within the world-system, this book joins scholarship such as De Loughry's *The Global Novel and Capitalism in Crisis* that centres a set of texts written in 'proximity to hegemony' (2020, p. 13). Again, the primary texts discussed in this book were produced by, and in most cases for, the people situated within this core. In addition to this, many of the fictions considered in this book were not produced simply in proximity to the core, *but emerge out of the very institutions that support and enable the core*. My point here (developed in more detail throughout the book) is that American Climate Emergency films such as Francis Lawrence's *I Am Legend* (2009) and Gareth Edward's *Godzilla* (2014), were produced in close collaboration with the network of military, capitalist and state interests that make the core function as core, and that are tasked with securitizing this core.

By focusing on narratives produced in the US, this book thus centres on a different experience of the capitalist world-ecology than that focused on by Niblett and Duncan. Their work convincingly shows that writing in the semiperipheries and peripheries registers capitalism as a force exerting considerable extractive, social, and ecological violence. Because of neoliberal deregulation and the long history of structural racism, the core in the US is, of course, closely stratified and heterogeneous, with the inner-city or rural poor living in close proximity to affluent suburbanites (to simplify a very complex social world). This means that such a registering of capitalism as an extractive and violent engine also occurs in core spaces.[7] However, and importantly, narratives from the privileged segment of the core still tend to register capitalism via the freedoms and comforts that it generates for the core. As De Loughry observes, the 'baseline of experiences' in the core, 'is that of the capitalist world-system, and more

---

[7] Deckard and Shapiro (2019) have argued that this transformation of the (American) core is now so pronounced that it makes sense to speak about the Neoliberal World-System.

specifically of a friction-free mobility during a period of relative financial and hegemonic stability from the 1990s to the early 2000s' (2020, p. 18). In this way, writing from the core is often about the pleasures of the automobile, the aircraft, the suburban bungalow, the shopping malls, the family constellations, and the casual love affairs that capitalism affords certain strata. In texts such as *Top Gun* (Scott 1986) or *Sex in the City* (1998–2010), capitalism is thus registered as an invisible but ubiquitous engine of a variety of racialized and sexualized privileges, powers, securities, pleasures, and comforts, and of the particular masculine and feminine modes of being that the exercise of these privileges and pleasures affords.

The world-literature perspective thus makes it possible to read the American Climate Emergency Narrative as a kind of core utterance; a multi-vocal yet contained expression of how this particular segment of the world-ecology *registers*, but also seeks to *manage*, the unfolding epochal capitalist world-ecological crisis. As a type of text that registers such crisis and the possible demise not just of the hegemony of the US, but of the entire world-system, the American Climate Emergency Narrative is relentlessly and repetitively violent. This violence is often disturbing and, as the coming chapters show, it can be understood as a registering of how capitalism's systemic violence has brought on a terminal crisis. However, as fiction from the core, the American Climate Emergency Narrative *manages* these eruptions through its erasure of the people and the extra-human lives that have long suffered ecological and social injustice and that are thus the historical recipients of this violence, and also by proposing that the solution to the abject violence it rehearses is to turn to the very institutions that, as described above, enable the core. Again, the solution to world-ecological crisis that is proposed by the American Climate Emergency Narrative is to invest in more of the militarized capitalism that produced the crisis.

By doing so, the American Climate Emergency Narrative can be said to be involved in an attempt to securitize the very system that has made the capitalist world-ecology. The tendency of American fiction to participate in such as project has been theorized within the growing field of literary security studies. A special issue of *American Literary History* edited by David Watson (2016) describes how American literature, from the dawn of the republic to the post-9/11 moment, has been saturated by a 'logic of security' understood as 'a diverse, heterogeneous, and dynamic set of knowledges, assumptions, and techniques' (p. 665) that surface in relation to a number of concerns in American life and history, from the notion of ownership and enclosure, to precarity and surveillance in the neoliberal US. Watson warns against viewing 'US cultural expression as security mechanisms' (p. 665), as doing so might reproduce the potentially divisive discourses through which states identify and distribute otherness and belonging. Yet, as this book demonstrates, many texts that can be described as American Climate Emergency Narratives do function in this way. This is arguably part of the work that core texts do: they rehearse the tenets that normalize and thus secure the dominant world-system it dominates, even as they register the anxieties and the violence by which the system is maintained.

By approaching the American Climate Emergency Narrative as a kind of core securitization enunciation, it is easier to understand why this voice sounds so different when compared to other voices coming from peripheral or semiperipheral parts of the world-system. To read these fictions as core expressions thus affords an important critical distance; it urges the scholar to read with caution, always aware of the fact that these texts speak about ecology and emergency from a certain position of economic, gendered, sexualized, and racialized power. Reading the American Climate Emergency Narrative as writing from the core also makes it easier to see how this body of text often speaks in unison with other voices that emanate from the US core. These other voices include the champions of liberal Green Capitalism (Sullivan 2009) but also purveyors of far-right ecofascism (Moore and Roberts 2022). The point here is that while these voices are often at odds with each other, they all assume, like the American Climate Emergency Narrative, that capitalism is an inevitable system. In this way, these literary and extra-literary voices are informed by what Mark Fisher (2009) has called capitalist realism. As I will return to, this is a realism premised on the understanding that there is no alternative to capitalism, even when this system is perceived to cause ruin and suffering.

MATERIAL AND ORGANIZATION

As described above, this book's main focus is the US as the hegemonic core of the existing world-system, but it also considers writing produced in Britain during the early phase of settler colonialism and, in its concluding chapter, texts from the US and Canadian semiperipheries. The study is primarily concerned with media forms such as novels, short stories, and film, but it also considers games and, to a marginal extent, the graphic novel. In doing so, it notes that the types of culture that require very substantial funding are particularly closely tied to the economic conditions that exist at the core, and to the institutions that produce these conditions. Thus, as already mentioned and as discussed in more detail in the coming chapters, some of the fiction produced in the core is created not by individuals as much as by networks within the capitalist economy. Such networks consist of globally dispersed creative industries, venture capitalist firms, global media conglomerates, and, notably, the US Department of Defense. These networks are ultimately geared towards producing spectacular and absorbing cinematic and ludological narratives that generate profit that can be accumulated within the world-system, and that disseminate specific political content that steers individual actors in certain directions (see Lenoir and Caldwell 2018; Wark 2009).

To look more closely at the corpus of this study, Chapters 2 and 3, and the first half of Chapter 4 explore texts published between the early seventeenth century and the late twentieth century. These are historical accounts, novels, plays, and Hollywood films that describe but also contribute to the making of America as the hegemon of the world-system, and that thus lay the ideological and imaginative foundation for the contemporary American Climate Emergency Narrative. After having discussed these origin stories, I turn to texts that have typically been filed as climate fiction by critics and literary scholars. Several of the texts discussed in this part of the book have received attention from climate fiction scholars and can be considered canonical to the field. Some of these employ an exclusively realist register and have been celebrated by literary critics, and some, like the books that make up the Southern Reach Trilogy, are far more speculative. However, the notion of the American Climate Emergency Narrative affords a possibility to widen the scope of the investigation and to look at novels and films that, like many of the canonical works, take place in futures clearly transformed by (capitalogenic) socio-ecological breakdown and that thus register epochal capitalist emergency, but that employ (monstrous) allegory and symbolism to tell the story. Climate fiction scholarship has typically ignored these texts or mentioned them only in passing. This is because they are, as Mark Bould (2021) observes, the 'sort of things the "serious" and the "literary" might wish the seas would swallow' (p. 4). By not ignoring them, this book is able to draw a much more complete map of the entangled body of texts that make up the American Climate Emergency Narrative.

The book consists of eight chapters where Chapter 2: *Settler Capitalist Frontiers* and Chapter 3: *Fossil Fictions* explore the genealogy of the

Climate Emergency Narrative alongside the material history that both elevated the US to the core of the world-system and set the world on the road towards world-ecological crisis. These chapters thus show how writing from the early colonial period to the end of the twentieth century registers the extractive relationship between land and people that enabled the emergence of the capitalist world-ecology. Chapter 2 moves through two distinct historical and narrative stages where the first is the settler colonial text that, from the very beginning, narrated ecology as an emergency-inducing space that needs to be violently combatted and the second the white pro-slavery and abolitionist novels that described slavery and the plantation as central to the growing US economy, but that also registered these as drivers of environmental breakdown. Chapter 3 focuses on the specific opportunities and emergencies that were connected to the turn to coal and then to oil in the nineteenth- and early twentieth-century US. The first section considers writing from the core that celebrates the arrival and extraction of coal in the semiperipheries of the US while the second section discusses how the militarization of petro-energies in the US gave rise to what can be termed the Petrowar narrative. As the chapter demonstrates, this was a self-assured narrative that afforded new ways of observing the world, but as the twentieth century drew to a close, it became coloured by anxieties connected to unfolding socio-ecological breakdown and possible world-system transformation.

Chapter 4: *The Irradiated* discusses American post-WWII fiction that registers the enormous, but also notably 'cheap', violence that nuclear weapons enabled. The release of this violence during and after the war produced stories set in futures where the climate of the planet has clearly broken down as an effect of nuclear warfare and where the capitalist world-system is crumbling. In some of these stories, a planet damaged by nuclear violence and lingering radiation takes the form of a gigantic, monstrous, and uncontrollable security threat that demands the attention of the military. By telling such stories, the chapter argues, these narratives must be considered as the formal beginning of the fully formed American Climate Emergency Narrative. As such, these texts demonstrate an increasing awareness of how militarized capitalism has helped to erode conditions for human and non-human life on the planet, but they still refuse to imagine alternatives to the system they understand as having destroyed the world.

While the first three chapters of the book read the American Climate Emergency Narrative alongside the historical ascent of the US, Chapters 5, 6, and 7 centre on texts written and produced after the turn of the millennium, when it is becoming increasingly clear that the biosphere is in crisis, this while the US is beginning to lose hegemonic control of the world-system. Chapter 5: *Geopolitics* focuses on American Climate Emergency Narratives that describe how climate breakdown has created international geopolitical tension and conflict. These are narratives that show the US and other major powers such as China, India, and Russia leveraging their considerable military resources to compete over, and secure, vital natural resources, in the process of which they establish new command over, or lose, hegemony over the world-system. These texts thus register the fact that continued erosion of the biosphere may produce world wars in the future, but they see no alternative to such development. The questions these texts pose are ultimately how such future wars can be won by the American security apparatus, but they also investigate what might happen if other actors in the world-system take advantage of American failure.

Chapter 6: *The Displaced* turns to the question of climate migration and reads a series of texts that follow climate refugees as they cross, or are prevented from crossing, heavily guarded US national or state borders. The chapter shows how some American Climate Emergency Narratives employ allegory to cast the racialized climate refugee as a border-scaling monster, but it also reveals how even texts that seek to critique the racist ideology that informs extractive capitalist border-thinking focus on the future plights of the white and privileged. Chapter 7: *Ruins* discusses the depiction of thoroughly eroded, post-apocalyptic worlds where capitalism, the commodity frontiers that have always fed capital, and the economic conditions that make standing armies possible, have disappeared. Much of the fiction considered in this chapter describes a Hobbesian world devoid of the material comforts, securities, and privileges previously enjoyed by white, middle-class communities. Even so, as the chapter reveals, the heroes of these texts do their best to honour the violent social contract established by extractive and militarized capitalism.

Throughout these first six chapters, I focus on culture produced in the hegemonic core and I show how most of it is involved in a tacit attempt to sanitize or elide the destructive and exhaustive processes that have produced the ongoing socio-ecological breakdown while at the same time lending support to the attempt to securitize this breakdown into a national security matter that can be resolved through military intervention and via the introduction of various states of exception. In this way, this book is a study of ideologies that can be traced back to the emergence of the settler capitalist state. However, the book's repeated return to, and description of, the ideologies that give form to this writing, risks naturalizing them. To resist this potential effect, the concluding Chapter 8: *Fallout Futures* poses that the writing analysed in the book can be considered as a form of ideological and intellectual residue or pathogen released by the core and invading the American imagination. Just as nuclear fallout travels via the stratosphere to all parts of the planet, this ideological fallout, energized by neoliberal globalization, reaches across the globe where it is shifting human environmental concerns towards an understanding of the future as a personal, regional, and/or national security concern. The aim of the final chapter is to refocus the needs of the planet as an entangled, multispecies habitat in the process of being destroyed by militarized capitalism. This chapter thus turns to climate narratives from the peripheries and semiperipheries that exist in close proximity to the US core. These, the chapter shows, register capitalism's epochal crisis from a radically different position and, in doing so, perform important imaginative work that makes it possible to think outside of the intellectual boundaries that make capitalism seem inevitable.

With the help of these chapters, and via the concept of the American Climate Emergency Narrative, this book hopes to contribute to the already ongoing radical turn that is informing the scholarly engagement with film, literature, games, and other cultural forms that centre the needs of the planet and of humans in all parts of the world, rather than the wants of the system currently involved in eroding it or the desires of the people currently being privileged by these systems. Just as fiction makes storyworlds and thus sets the limits for what futures can be imagined, scholarship is involved in building worlds and futures for readers. This is an important realization for all scholarship and particularly for ecocritical interventions performed at a time when the Earth System may be approaching what climate scientists have termed tipping points (Dakos et al. 2019; Lenton et al. 2019). These tipping points are stages when the detrimental effects produced by global warming accelerate and cascade to such an extent that the niche that makes human life convenient rapidly erodes in many parts of the world. That said, and as argued by Mahanty et al. (2023), what should motivate a transformation of both the world-system and ecocritical scholarship is not necessarily just the prospect of a looming yet still future ecological threshold, but the socio-ecological rupture already caused by the inexorable slow violence of colonial capitalism.

The understanding that capitalism is driving ongoing socio-ecological breakdown prompts an activism very different from that produced by the insistence that the human species is the engine of the climate crisis. Rather than changing humanity, whatever that might mean, the obvious way forward is to profoundly alter the capitalist system that burns oil and coal, and that extracts minerals, human and extra-human life, and accumulates these as capital. To perform such systemic change is not an easy task. Capitalism is folded into most people's lives on a very deep level, and most humans have been conditioned since birth to become integrated with it in one way or another. Even so, it is a fundamentally *possible* task. Despite the chorus of voices (many emanating out of the fiction studied in this book) that insists that no other social, ecological, or economic world is possible than the one produced by capitalism, there are futures beyond it. Ecocritical scholarship that resists these voices to instead centre the capitalist histories, systems, responses, and people that are causing socio-ecological breakdown, can assist in creating the intellectual platforms that make it possible to actively and systematically work towards human social and biospheric renewal. This book ultimately hopes to participate in the construction of such platforms.

WORKS CITED

Agamben, Giorgio. 2005. *State of Exception*. Translated by Kevin Attell. Chicago: University of Chicago Press.

Althusser, Louis. 2014. *On the Reproduction of Capitalism: Ideology and Ideological State Apparatuses*. London: Verso.

Andersen, Gregers. 2020. *Climate Fiction and Cultural Analysis: A New Perspective on Life in the Anthropocene*. Oxon: Routledge.

Arrighi, Giovanni. 1978. 'Towards a Theory of Capitalist Crisis.' *New Left Review* 111 (3): 3–24.

Arrighi, Giovanni. 1994. *The Long Twentieth Century: Money, Power, and the Origins of Our Times*. London: Verso.

Birkmann, Joern, Ali Jamshed, Joanna M McMillan, Daniel Feldmeyer, Edmond Totin, William Solecki, Zelina Zaiton Ibrahim, Debra Roberts, Rachel Bezner

Kerr, and Hans-Otto Poertner. 2022. 'Understanding Human Vulnerability to Climate Change: A Global Perspective on Index Validation for Adaptation Planning.' *Science of The Total Environment* 803: 150065.

Bould, Mark. 2021. *The Anthropocene Unconscious: Climate Catastrophe Culture.* London: Verso.

Ceballos, Gerardo, Paul R. Ehrlich, Anthony D. Barnosky, Andrés García, Robert M. Pringle, and Todd M. Palmer. 2015. 'Accelerated Modern Human-Induced Species Losses: Entering the Sixth Mass Extinction.' *Science Advances* 1 (5): e1400253.

Cedamia. 2022. 'Are Climate Emergency Declarations still happening?' https://www.cedamia.org/climate-emergency-declarations/are-climate-emergency-declarations-still-happening/. Accessed January 12, 2024.

Crutzen, Paul, and Eugene Stoermer. 2000. 'The "Anthropocene".' *IGBP Newsletter* 41: 17–18.

Dakos, Vasilis, Blake Matthews, Andrew P Hendry, Jonathan Levine, Nicolas Loeuille, Jon Norberg, Patrik Nosil, Marten Scheffer, and Luc De Meester. 2019. 'Ecosystem Tipping Points in an Evolving World.' *Nature Ecology & Evolution* 3 (3): 355–362.

Davis, Heather, and Zoe Todd. 2017. 'On the Importance of a Date, or, Decolonizing the Anthropocene.' *ACME: An International Journal for Critical Geographies* 16 (4): 761–780.

De Loughry, Treasa. 2020. *The Global Novel and Capitalism in Crisis.* Chamalthusser: Palgrave Macmillan.

Deckard, Sharae. 2019. 'Trains, Stone, and Energetics: African Resource Culture and the Neoliberal World-Ecology.' In *World Literature, Neoliberalism, and the Culture of Discontent*, edited by Sharae Deckard and Stephen Shapiro, 239–262. Chamalthusser: Palgrave Macmillan.

Deckard, Sharae, and Stephen Shapiro. 2019. *World Literature, Neoliberalism, and the Culture of Discontent.* Chamalthusser: Palgrave Macmillan.

Denoon, Donald. 1983. *Settler Capitalism: The Dynamics of Dependent Development in the Southern Hemisphere.* New York: Oxford University Press.

Dixson-Declève, Sandrine, Owen Gaffney, Jayati Ghosh, Jørgen Randers, Johan Rockström, and Per Espen Stocknes. 2022. *Earth for All: A Survival Guide for Humanity: A Report to the Club of Rome.* Gabriola Island: New Society Publishers.

Duncan, Rebecca. 2022. 'Gothic in the Capitalocene: Imagining World-Ecological Crisis from the South African Postcolony.' In *Dark Scenes from Damaged Earth: The Gothic Anthropocene*, edited by Justin D. Edwards, Rune Graulund, and Johan Höglund, 175–194. Minneapolis: Minnesota University Press.

Duncan, Rebecca, Mike Classon Frangos, Eleonor Marcussen, and Emily Hanscam. 2023. 'Emergency.' *Environment and History* 29 (4): 476–482. https://doi.org/10.3197/096734023X16945097374245.

Edwards, Gareth, director. 2014. *Godzilla*. Warner Bros. Pictures.

Fisher, Mark. 2009. *Capitalist Realism: Is There No Alternative?* Ropley, Hants: John Hunt Publishing.

Foster, John Bellamy. 1999. 'Marx's Theory of Metabolic Rift: Classical Foundations for Environmental Sociology.' *American Journal of Sociology* 105 (2): 366–405.

———. 2002. *Ecology against Capitalism*. New York: New York University Press.

Foster, John Bellamy, Brett Clark, and Hannah Holleman. 2021. 'Marx and the Commons.' *Social Research: An International Quarterly* 88 (1): 1–30.

Foucault, Michel. 1991. *The Foucault Effect: Studies in Governmentality*. Chicago: University of Chicago Press.

Gómez-Barris, Macarena. 2017. *The Extractive Zone*. Durham, NC: Duke University Press.

Holleman, Hannah. 2018. *Dust Bowls of Empire*. New Haven, CT: Yale University Press.

Hulme, Mike. 2019. 'Climate Emergency Politics Is Dangerous.' *Issues in Science and Technology* 36 (1): 23–25.

Iossifidis, Miranda Jeanne Marie, and Lisa Garforth. 2022. 'Reimagining Climate Futures: Reading Annihilation.' *Geoforum* 137: 248–257.

IPCC, 2022. *Climate Change 2022: Impacts, Adaptation and Vulnerability*, edited by Hans O Pörtner, Debra C Roberts, Helen Adams, Carolina Adler, Paulina Aldunce, Elham Ali, Rawshan Ara Begum, Richard Betts, Rachel Bezner Kerr, and Robbert Biesbroek.

James, Erin. 2015. *The Storyworld Accord: Econarratology and Postcolonial Narratives*. Lincoln: University of Nebraska Press.

Jameson, Fredric. 1981. *The Political Unconscious: Narrative as a Socially Symbolic Act*. Ithaca, NY: Cornell University Press.

Klein, Naomi. 2007. *The Shock Doctrine: The Rise of Disaster Capitalism*. New York: Picador.

———. 2015. *This Changes Everything: Capitalism Vs. The Climate*. New York: Simon and Schuster.

Koch, Alexander, Chris Brierley, Mark M Maslin, and Simon L Lewis. 2019. 'Earth System Impacts of the European Arrival and Great Dying in the Americas after 1492.' *Quaternary Science Reviews* 207: 13–36.

Kolbert, Elizabeth. 2014. *The Sixth Extinction: An Unnatural History*. New York: Henry Holt.

Lawrence, Francis, director. 2009. *I Am Legend*. Warner Bros. Pictures.

Lazarus, Neil. 2011. *The Postcolonial Unconscious*. Cambridge: Cambridge University Press.

Lenoir, Timothy, and Luke Caldwell. 2018. *The Military-Entertainment Complex*. Cambridge, MA: Harvard University Press.

Lenton, Timothy M, Johan Rockström, Owen Gaffney, Stefan Rahmstorf, Katherine Richardson, Will Steffen, and Hans Joachim Schellnhuber. 2019. 'Climate Tipping Points—Too Risky to Bet Against.' *Nature* 575: 592–595.

Lewis, Simon L, and Mark A Maslin. 2015. 'Defining the Anthropocene.' *Nature* 519 (7542): 171–180.

Luxemburg, Rosa. 2015. *The Accumulation of Capital*. Abingdon: Routledge. 1913.

Mahanty, Sango, Sarah Milne, Keith Barney, Wolfram Dressler, Philip Hirsch, and Phuc Xuan To. 2023. 'Rupture: Towards a Critical, Emplaced, and Experiential View of Nature-Society Crisis.' *Dialogues in Human Geography* 13(2): 177–196.

Malm, Andreas. 2016. *Fossil Capital: The Rise of Steam Power and the Roots of Global Warming*. London: Verso.

———. 2018. *The Progress of This Storm: Nature and Society in a Warming World*. London: Verso.

Malm, Andreas, and Alf Hornborg. 2014. 'The Geology of Mankind? A Critique of the Anthropocene Narrative.' *The Anthropocene Review* 1 (1): 62–69.

Malpas, Imogen. 2021. 'Climate Fiction Is a Vital Tool for Producing Better Planetary Futures.' *The Lancet Planetary Health* 5 (1): e12–e13.

Marzec, Robert P. 2002. 'Enclosures, Colonization, and the Robinson Crusoe Syndrome: A Genealogy of Land in a Global Context.' *Boundary* 29 (2): 129–156.

———. 2015. *Militarizing the Environment: Climate Change and the Security State*. Minneapolis: University of Minnesota Press.

Moore, Jason W. 2015. *Capitalism in the Web of Life: Ecology and the Accumulation of Capital*. New York: Verso.

———. 2016. *Anthropocene or Capitalocene? Nature, History, and the Crisis of Capitalism*, edited by Jason W. Moore. Oakland: PM Press.

Moore, Sam, and Alex Roberts. 2022. *The Rise of Ecofascism: Climate Change and the Far Right*. Cambridge: Polity.

Moretti, Franco. 1996. *Modern Epic: The World System from Goethe to García Márquez*. London: Verso.

———. 2000. 'Conjectures on World Literature.' *New Left Review* 1: 54–68.

Neeson, Jeanette M. 1993. *Commoners: Common Right, Enclosure and Social Change in England, 1700–1820*. Cambridge: Cambridge University Press.

Niblett, Michael. 2012. 'World-Economy, World-Ecology, World Literature.' *Green Letters* 16 (1): 15–30.

———. 2020. *World Literature and Ecology: The Aesthetics of Commodity Frontiers, 1890–1950*. Chamalthusser: Palgrave Macmillan.

Nixon, Rob. 2011. *Slow Violence and the Environmentalism of the Poor*. Cambridge, MA: Harvard University Press.

Patel, Raj, and Jason W Moore. 2018. *A History of the World in Seven Cheap Things: A Guide to Capitalism, Nature, and the Future of the Planet*. London: Verso.

Parry, Benita. 2004. *Postcolonial Studies: A Materialist Critique*. Abingdon: Routledge.

Quijano, Anibal. 2000. 'Coloniality of Power and Eurocentrism in Latin America.' *International Sociology* 15 (2): 215–232.

Quijano, Anibal, and Immanuel Wallerstein. 2000. 'Coloniality of Power and Eurocentrism in Latin America.' *International Sociology* 15 (2): 215–232.

Robinson, William I. 2020. *The Global Police State*. London: Pluto Press.

Sánchez Prado, Ignacio M. 2019. 'Mont Neoliberal Periodization: The Mexican "Democratic Transition," from Austrian Libertarianism to the "War on Drugs".' In *World Literature, Neoliberalism, and the Culture of Discontent*, edited by Sharae Deckard and Stephen Shapiro 93–110. Chamalthusser: Palgrave Macmillan.

Scott, Tony, director. 1986. *Top Gun*. Paramount Pictures.

Shapiro, Stephen, and Philip Barnard. 2017. *Pentecostal Modernism: Lovecraft, Los Angeles, and World-Systems Culture*. London: Bloomsbury Publishing.

Sullivan, Sian. 2009. 'Green Capitalism, and the Cultural Poverty of Constructing Nature as Service-Provider.' *Radical Anthropology* 3: 18–27.

Sultana, Farhana. 2022. 'The Unbearable Heaviness of Climate Coloniality.' *Political Geography* 99: 102638.

Trexler, Adam. 2015. *Anthropocene Fictions: The Novel in a Time of Climate Change*. Charlottesville: University of Virginia Press.

VanderMeer, Jeff. 2014a. *Annihilation: A Novel*. New York: Farrar, Straus and Giroux.

———. 2014b. *Authority*. New York: Farrar, Straus and Giroux.

———. 2014c. *Acceptance*. New York: Farrar, Straus and Giroux.

Vitale, Alex S. 2017. *The End of Policing*. London: Verso.

Von Mossner, Alexa Weik. 2017. 'Sensing the Heat: Weather, Water, and Vulnerabilities in Paolo Bacigalupi's *The Water Knife*.' In *Real-Yearbook of Research in English and American Literature: Vol. 33 (2017): Meteorologies of Modernity. Weather and Climate Discourses in the Anthropocene*, edited by Sarah Fekadu, Hanna Straß-Senol, and Tobias Döring, vol 33, 173–190. Tübingen: Narr Francke Attempto Verlag.

Wallerstein, Immanuel M. 2004. *World-Systems Analysis: An Introduction*. Durham: Duke University Press.

Wark, McKenzie. 2009. *Gamer Theory*. Cambridge: Harvard University Press.

Warwick Research Collective (WReC). 2015. *Combined and Uneven Development: Towards a New Theory of World-Literature.* Liverpool: Liverpool University Press.

Watson, David. 2016. 'Introduction: Security Studies and American Literary History.' *American Literary History* 28 (4): 663–676.

Wenzel, Jennifer. 2019. *The Disposition of Nature: Environmental Crisis and World Literature.* New York: Fordham University Press.

Yelling, James Alfred. 1977. *Common Field and Enclosure in England 1450–1850.* Basingstoke: Macmillan Press.

Yusoff, Kathryn. 2018. *A Billion Black Anthropocenes or None.* Minneapolis: University of Minnesota Press.

CHAPTER 2

Settler Capitalist Frontiers

THE ORIGINS OF THE CLIMATE EMERGENCY NARRATIVE

The origins of the Climate Emergency Narrative cannot be grasped or properly analysed if the origins of the climate emergency, or of socio-ecological breakdown, are not clearly understood. To revisit an important point made in the introduction, there is currently disagreement on what began the ongoing and profoundly interrelated social and biospheric crisis. In 2016, atmospheric chemists Will Steffen and Paul J. Crutzen joined American environmental historian John R. McNeill to propose, in the article 'The Anthropocene: Are Humans Now Overwhelming the Great Forces of Nature?', that the beginning of the Anthropocene was the discovery of fire by Homo Erectus 'a couple of million years ago' (2016, p. 614). This is a proposition that makes the current crisis seem like something inevitable, a result of the evolution of the human species. By contrast, and as discussed in more detail in Chapter 4, the Anthropocene Working Group (focusing on the geological stratigraphic record) has suggested that what it terms the Anthropocene should be located 'historically at the moment of detonation of the Trinity A-bomb at Alamogordo in 1945' (Zalasiewicz et al. 2015). This is a proposition that quite shockingly erases the long world-system history that arguably produced WWII and the technology of the atomic bomb (as well as the willingness to use it on Hiroshima and Nagasaki).

© The Author(s) 2024
J. Höglund, *The American Climate Emergency Narrative*,
New Comparisons in World Literature,
https://doi.org/10.1007/978-3-031-60645-8_2

As explained in the introduction, this book joins environmental historians and sociologists such as John Bellamy Foster (1999, 2002), Jason W. Moore (2015), Heather Davis and Zoe Todd (2017), Macarena Gomez-Barris (2017), Hannah Holleman (2018), and many others that argue that the origins of the biospheric crisis are systemic and capitalogenic rather than evolutionary and species related, and that resist the notion that the drivers of the crisis arrived fully formed in 1945 in the shape of an atomic bomb. Following their work, I locate the beginning of the ongoing biospheric crisis to the beginning of capitalism and colonialism and the emergence of the capitalist world-ecology. The region known as North America played a key role in this development. This space was colonized by several European nation-states in the late sixteenth and early seventeenth century and treated by them as a commodity periphery. Over the years, these colonies merged and grew to become, first under the auspices of the British Empire, then as a new nation-state, a semiperipheral, but increasingly central, region within a world-system ruled by the British hegemonic core. Importantly, many of the erosive technologies (settler colonialism, chattel slavery, the monocrop planation, the reservation, and the combustion engine) that are central to the world-system today and that have contributed to the present socio-ecological crisis were invented or honed within this region.

This understanding of the origins of the socio-ecological breakdown makes it possible to collate the beginnings of the present crisis with the texts produced by those privileged by this development. This chapter thus traces the genealogy of the American Climate Emergency Narrative back to the time when European actors began to settle and enclose the North American continent, eliminating Indigenous people and establishing the extractive plantation paradigm. This development was often notoriously violent, but as the chapter shows, it was described as a necessary and inevitable response to *emergency* in texts from the period. It furthermore demonstrates how some early texts registered the fluctuating violence and privileges, and the booms and crises, that accompanied the first stages of America's transformation from commodity periphery to the seat of a new world-system hegemon. By doing so, the chapter reveals how these early narratives were obsessed with resources (clean water, fertile soil, teeming oceans, various types of human, and mineral energies), and how even texts that registered the violence of colonial capitalism perceived it as inevitable. As the coming chapters demonstrate, such obsessions and such securitization logics are central to the American Climate Emergency Narrative.

Because this book focuses on the dominant core narrative, this chapter discusses texts produced by and for those who helped engineer the early capitalist and colonial project in North America. Thus, the chapter first returns to the beginnings of the world-ecology and settler colonialism, to see how one of the first agents of this emerging development described the encounter between British settlers and the land and the Indigenous people living on it and with it. The chapter then explores texts written by settler authors in antebellum US, who were concerned with the establishment and maintenance of the slave-operated plantation. The chapter shows how these two stages—settlement and the operation of the slave-driven plantation—produced writing profoundly preoccupied with the securitization of extractable land and thus with the *commodity frontiers* that were established by settlers. As Raj Patel and Jason W. Moore (2018) have argued, capitalism 'not only *has* frontiers; it exists only *through* frontiers, expanding from one place to the next, transforming socioecological relations, producing more and more kinds of goods and services that circulate through an expanding series of exchanges' (p. 19, italics in the original). Furthermore, as Moore and Patel also note, the effort to establish and monopolize these frontiers, to seize the land on which they depend, and to secure them against competition from other settlers, typically required violence.[1] This violence was frequently so extensive that the establishment of many commodity frontiers has been historicized as *wars* by scholarship. The history of American commodity frontiers has thus been written as a series of range wars (Drago, 1970), sheep wars (Carlson, 1984), tobacco wars (Campbell, 2005), coal wars (Shogan, 2004), railroad wars (Starr, 2012), oil wars (Williams, 2020; Winegard, 2016), and so on. These wars were partly economic, but they were also actual military confrontations. As I discuss in more detail in Chapter 3, the armed conflict between unionizing coal labourers and anti-union (private and state) forces in West Virginia in 1921 was the largest battle fought on US soil since the Civil War (Shogan, 2004).

---

[1] Similar points have been made by Indigenous, postcolonial, decolonial, and Marxist scholars who argue that the acquisition of land and the securitization of extraction on the commodity frontier were very violent undertakings. As Patrick Wolfe (2006) has influentially observed, settler colonialism works according to a 'logic of elimination' that encourages both genocide and cultural erasure.

By focusing on the centrality of the commodity frontier and the settler capitalist transformation of land into what Moore (2015) calls Cheap Nature,[2] the chapter provides the American Climate Emergency Narrative with a genealogy that demonstrates how the origins of this narrative are closely intertwined with the transformation of the US into the hegemon of the world-system. Because the chapter engages with a very long material, economic, military, and cultural history, it is by necessity cursory. It thus references and combines, rather than contributes to, previous scholarship by economic, environmental, and literary historians. What it adds to this work is a perspective through which it becomes possible to see how the texts discussed write human and extra-human nature both as repositories of enormous wealth and as entities that must be made secure even when such securitization involves the application of massive violence. This way of narrating ecology, the chapter shows, characterizes both writing that depicts the early settler colonial endeavour, and later texts about the slave-operated monocrop plantation. It is present, as the chapter reveals, even in anomalous passages where a certain, politically unconscious awareness of the destructive nature of capitalism surfaces; where human lives and ecologies are described as depleted and violated. From the vantage of the genealogy that the chapter establishes, the twenty-first-century American Climate Emergency Narrative's relative disinterest in the climate and its preoccupation with substantive violence, ownership, crisis, and (lost) privilege becomes easier to understand. It is also when this narrative is perceived as the most recent incarnation of a tradition that has always been concerned with the need to manage ecological/human crises and emergencies that its peculiar registering of epochal crisis makes the most sense.

---

[2] As discussed in the introduction to this book, and as described by Jason W. Moore (2015), Cheap Nature is nature that capitalism can commodify with relative ease and thus at a low cost. As Moore observes, 'capitalism has survived not by destroying nature (whatever this might mean), but through projects that compel nature-as-oikeios to work harder and harder—for free, or at a very low cost' (p. 23). Capitalism thus thrives when there is an abundance of Cheap Nature and the imperative is to move away from commodity frontiers where nature that can be sourced cheaply is running out, to new frontiers where it can again be found. The current problem (for capitalism and for the planet it has reshaped is that 'it is becoming increasingly difficult to get nature—of any kind—to work harder' (p. 23).

The Settler Capitalist Text

Moore locates the formal beginning of what he terms the Capitalocene to the 1450s, as it was at this time that 'an epochal shift in the scale, speed, and scope of landscape transformation across the geographical expanse of early capitalism occurred' (2016, p. 96). This epochal shift was energized by the highly fertile soil that existed in America and that provided the settlers with an opportunity to extract nature cheaply. It took a few centuries before the reorganization of ecology in America (and in other parts of the world colonized by Europeans) resulted in the actual warming of the planet, but the arrival of the first European settlers still had a noticeable effect on the climate of the Earth System. The viruses and bacteria that travelled with the first colonists brought on what has been described as the 'Great Dying', an event that reduced the Indigenous population by an estimated 90 per cent, from approximately 60 million people to 5 million (Koch et al., 2019). In the wake of this dying, the agriculture and the animal husbandry on which Indigenous people in America had subsided partially collapsed. When fields and farms were abandoned, massive reforestation occurred. This, as several studies have suggested, bound more atmospheric CO_2, and contributed to a lowering of the temperature of the entire planet (Kaplan et al., 2011; Koch et al., 2019; Nevle & Bird, 2008). Geoffrey Parker (2013) has argued that this drop in temperature significantly contributed, alongside reduced solar radiation and increasing volcanic activity in other parts of the world, to what has been termed the Little Ice Age which lasted from the early fourteenth century to the middle of the nineteenth. In this way, the arrival of European settler capitalism can be said to have affected the Earth System at a very early stage.[3]

As this and coming chapters illustrate, the extractive understanding of land was expressed in a number of official and unofficial texts. These texts mediated and operationalized this understanding, and they also remain

---

[3] Simon L. Lewis and Mark A. Maslin (2015) have influentially argued that this dramatic and sustained drop in temperature should be considered as the first evidence that human *systemic* activity is determining the progress of the Earth System. They thus propose that the drop in temperature that the dying of Indigenous people and the reforestation of America produced should be considered clear evidence of a global geological shift or, in geologic terms, a so-called 'golden spike'. Because the temperature drop was global, Lewis and Maslin have suggested the term Orbis (World) Spike to name this global geologic event.

as testimonies of it. An important early stage in this development was the writing of a series of charters that opened America to British settler colonialism. The first was the Virginia Charter of 1606 which enclosed a portion of land on the American East Coast and leased it to the Virginia Company of London. Empowered by this charter, the company invested in ships, supplies, and manpower and established the Jamestown settlement. In 1628, the *Charter of the Massachusetts Bay Company*, signed by Charles I, opened the region which is known today as New England to colonization. Like the Virginia Charter, this outlined the geographical borders of the colony and also imagined the commodity frontiers that could be assumed to exist in this land and that the company was instructed to exploit. Thus, the Massachusetts Charter bequeathed the Massachusetts Bay Company the 'firm lands, soils, grounds, havens, ports, rivers, waters, fishing, mines' of this territory and asked the company to extract, on behalf of itself and the Crown, 'minerals, as well royal mines of gold and silver, as other mines and minerals, precious stones, quarries, and all and singular other commodities' (Charter of the Massachusetts Bay Company cited by Anon, 1814, pp. 18–20).

While the desire to create a haven for a particular type of religious practice may have been what encouraged many of the settlers to make the journey to America, it was the hope of discovering such commodities that funded the journey.[4] The extractive work performed on these new frontiers also paid for the wars that were fought to create, protect, and expand colonial settlements. As postcolonial and world-literature scholarship has demonstrated, the violence used by extractive colonial capitalism to open frontiers up was experienced and registered in drastically different ways by people living in different parts of the emerging world-system. Among Indigenous people in the periphery, this violence was predictably registered as abject, grotesque, and horrific (Acebo, 2021; Brandt, 2017). However, in the emerging core, the same violence was frequently sanitized and narrated as essential to the survival of the early settler colonial/ capitalist endeavour.

One of the most striking examples of how human/ecological violence was produced, narrated, and rationalized in America during the early colonial period is Captain John Underhill's *Newes from America; Or, A New*

---

[4] See J. F. Martin's *Profits in the Wilderness: Entrepreneurship and the Founding of New England Towns in the Seventeenth Century* (1991) and Agnès Delahaye, 'Managing New England' (2020).

and Experimentall Discoverie of New England, published in London in 1638. Underhill was one of several officers recruited by the Massachusetts Bay Company to train the colony's militia and to help keep the new settlement secure. In 1635, epidemics and unusually bad weather appear to have created food shortages that strained the resources of both settlers and Indigenous people (Grandjean, 2011). When tensions between the settlers and the Pequot nation—one of the many Indigenous peoples in the area—erupted into open conflict, Underhill was one of the officers tasked with managing the situation. *Newes from America* is his account of the event that followed.

The first thing to note is that while *Newes from America* exists within material and cultural history as a text that reports events in the American periphery, it was written in and for a growing European capitalist society. As Giovanni Arrighi argues in *The Long Twentieth Century* (1994), the early seventeenth century is a sequence within the capitalist world-system's second Dutch cycle. However, Britain was already creating the conditions that would allow the nation to replace the Dutch as the hegemon of this system in the latter half of the eighteenth century. It is from this vantage that *Newes from America* maps New England as a commodity frontier that may yield significant, commodifiable resources. Underhill thus describes 'Augumeaticus' (in what is today northern Maine) as 'a place of good accommodation [...] worthy to bee inhabited, a soyle that beares good corne, all sorts of graine, flax, hemp, the Countrey generally will afford' (p. 17). Similarly, he argues that in 'Puscataway [...] there was growne in the last yeare [...] as good English graine as can grow in any part of the world' (p. 18). In this way, the 'newes' that Underhill conveys is that New England is a territory rich in Cheap Nature and that this territory is ripe for enclosure and extraction. The spaces he describes can be turned into commodity frontiers where certain staples and energies can be sourced for the benefit of the colonial/capitalist system he is involved in establishing.

At the same time, the book is a candid and often disturbing description of the war fought between Puritan settlers and the Indigenous Pequot. This was a long-lasting conflict that involved several battles before culminating in what has been described as an almost complete genocide of the Pequot nation (Cave 1996). The final confrontation between the Pequot (already reduced by decades of illness) and the settlers was the massacre at the fortified Pequot city Mystic where, as Underhill writes 'foure hundred [...] men, women, and children' (p. 35) were murdered. For the small number of Pequot that survived, exile, '[c]aptivity, slavery, and death' (Grandjean 2011, p. 99) followed.[5]

Notably, the war that Underhill waged was aimed not only at Indigenous people but also at the very land that the Pequot lived on. Underhill thus repeatedly describes how he and his men 'burnt and spoyled both houses and corne in great abundance [...] burnt their houses, cut downe their corne, destroyed some of their dogges in stead of men' (p. 6). When Underhill cannot find Pequots to kill, he will spend 'the day burning and spoyling the Countrey' (p. 14). In this way, the destruction of Indigenous land is also, in this text, a way of *managing* the land. To make the land safe for extraction, and to harness its potential as a settler capitalist utopia, it must first be transformed through extensive, military violence.

Underhill is aware that the violence he and his men perform on both people and land is excessive. He even inserts a question that he apparently has been asked, and that some of his readers may also be posing, into his narrative: 'Why should you be so furious (as some have said) should not Christians have more mercy and compassion?' (p. 35). Underhill's explanation is that this violence is a response to an emergency, and to the acute dearth of security this emergency is producing. He explains that 'our people were so farre disturbed, and affrighted with their [the Pequot's] boldnesse that they scarce durst rest in their beds: threatning persons and cattell to take them, as indeed they did' (p. 20). The (genocidal) violence that he and his men perform is thus cast as the necessary response to this threat; a way to make the settlement safe. The need to secure life and property, the order enforced by the enclosure, thus takes priority over the Christian norm of compassion. It is thanks to this violence that the fertile soil and bountiful geography he travels through has 'fallen into the hands of the English' (p. 2). This is also part of the 'newes' that Underhill spreads: America is a territory being primed for enclosure and extraction, and to acquire it, it is necessary to kill Indigenous people and to do violence on the land that belongs to them. This is how you secure the land for extraction.

If the American Climate Emergency Narrative is understood as a body of texts that promotes securitization regimes and military violence as a resolution to social and ecological crisis, as this book argues, *Newes from America* marks one important instance where this narrative emerges. In

---

[5] David R. Wagner and Jack Dempsey (2004) have challenged this history in *Mystic Fiasco: How the Indians Won the Pequot War*.

Newes from America, the environment appears as both a repository of abundant Cheap Nature and as a potentially hostile, insurgent space. To take advantage of its potential for early capitalism, to make it 'fit' for '*New Plantations*' (Underhill p. 1, italics in the original), the land must be violently cleansed before being reorganized. The killing and destruction of both people and land that Underhill's militia engages in, is imagined as the first and necessary step towards such reorganization.

Richards Slotkin has influentially argued that in the American novel, 'the wilderness must be destroyed so that it can be made safe for the white women and the civilisation she represents' (1973, p. 554). More recently, Kate Holterhoff (2019) has considered the same issue in a study that reveals how a great deal of English and American popular fiction from the nineteenth century narrated the historical change caused by European expansion via 'the genocidal extermination of native populations' (p. 279). As Holterhoff observes, this genocidal violence is often displaced so that the destruction of Indigenous people is told via the killing of charismatic megafauna such as whales, mammoths, or cave-bears who inhabit, or once inhabited, the wilderness. Written before the need to 'regenerate' America was perceived, *Newes from America* can be read as a precursor to this type of narrative. In Underhill's text, Indigenous people and the land they inhabit are destroyed to make way not simply for white settlers, but for a new *relationship* to the planet. Indeed, what is destroyed in *Newes from America* is not (obviously) ecology as such, but a certain kind of understanding of the relationship between nature and human society. Thus, the 'newes' that *Newes from America* ultimately conveys is that unenclosed ecologies are unruly and dangerous, that they host subversive entities such as Indigenous people who, by resisting settlement, oppose the very tenets that inform early settler capitalist society: the right to enclose and privatize the land, to extract commodifiable energies from this land, and to accumulate the surplus that results from this process.

The Plantation Text

The material and epistemological violence practised and rehearsed by the early settler community in America made it possible to establish the new ecological system that Underhill calls 'Plantation'. This can be described as a regime through which 'labor power' and 'extra human nature' (Moore 2015, p. 105) were put to work to exploit various agrarian commodity frontiers—tobacco, rice, indigo, sugar cane, cotton— as cheaply as possible. In this way, the plantation needs to be understood not primarily as a farm where food for the local community was grown (although such farms certainly existed), but as an institution that serviced the world-system with cash crops that could be refined and sold on the world market. In this way, the plantation introduced what Lisa Tilley has termed 'a strange industrial order' (2020) in America, strange because of its 'rationalized order, its technologized form, the way it materializes mastery over nature, and, ultimately, the way it destroys ecologies in the name of 'improvement'' (p. 67). In other words, beginning with the first sugar plantations, the plantation introduced a new type of industrialized agrarian paradigm, where specific energies were effectively sourced for a growing world market.

As Tilley argues, the initial strangeness of the plantation has been elided in late capitalist modernity so that it is now 'a thoroughly normalized landscape form' (p. 67). Indeed, the monocrop plantation and the cattle or poultry farm where rice, wheat, soy, pine, palm oil, and meat are sourced for a world economy, have long been the dominant form of agriculture. In an effort to recognize how the plantation has shaped the planet and, by doing so, contributed to biospheric breakdown, Donna Haraway (2015, 2016) has offered the concept of the Plantationocene as an alternative or complementary denominator to the Capitalocene. In Haraway's definition, the Plantationocene names the 'devastating transformation of diverse kinds of human-tended farms, pastures, and forests into extractive and enclosed plantations, relying on slave labor and other forms of exploited, alienated, and usually spatially transported labor' (2015, p. 162, n. 5). Natalie Aikens et al. (2019), Janae Davis et al. (2019), Katherine Yusoff (2018), and Wendy Wolford (2021) have similarly focused on the plantation as an extractive, violent, and fundamentally racialized system that has turned much of the Global South into land sourced for the commodities requested in the Global North. As Hanna Hollman has observed in *Dust Bowls of Empire: Imperialism, Environmental Politics, and the Injustice of "Green" Capitalism* (2018), the need to source these commodities as cheaply as possible is exhausting the soil to such an extent that farmland in many parts of the planet is turning into arid dust.

Just like the elimination of Indigenous people demanded a certain kind of violence, the establishment and maintenance of the plantation required a particular set of violent strategies. The pacification of enslaved people on the plantation, and the trans-Atlantic and internal slave trade that serviced the plantation, were inherently violent, and just like the genocide of Indigenous people, this violence was cast as a security measure vital to the plantation system itself. In fact, it can be argued that the plantation introduced a perpetual state of exception as defined by Giorgio Agamben (2005). Understood as such, the plantation can be considered a permanent response to the emergency that Indigenous people, Indigenous land, and enslaved people were seen to constitute by settlers. In the words of David Lloyd (2012), 'the state of exception and the excluded inside have been constitutive elements of colonialism, and in particular of settler colonial formations, from the Indian reservations [...] to slave plantations and their precociously modern architectures' (p. 72).[6] In this way, and to summarize the points made above, the plantation was central to the organization, extraction, and depletion of both human labour and extra-human nature. At the same time, the plantation was a configuration designed to manage the internal pressures of the world-system: the imperative to feed commodities to the market and to manage the cyclical crises that are, as argued by Arrighi (1978) and Moore (2011), endemic to capitalism.

These perceived insecurities, and the violence employed by military or paramilitary agents on behalf of the capitalist, colonial endeavour, were narrated and mediated by fiction written by people located within the fiercely stratified US semiperiphery as it existed during much of the nineteenth century. As Wendy Wolford (2021) has argued, the plantation not only 'propelled colonial exploration, sustained an elite, perpetuated a core–periphery dualism within and between countries', but it also 'shaped both the cultures we consume and the cultural norms we inhabit and perform' (1623). In this way, as proposed by Aikens et al. (2019), 'the plantation is as much a narrative construction as it is a literal place, one that always underlies political debates about national belonging, borders, labor, resources, ethnicity, and race' (np).

The texts that most clearly register the violence done to land and people on the plantation were not written at the core, of course, but are what scholarship has termed slave narratives. The most widely circulated at the time include Olaudah Equiano's *The Interesting Narrative of the Life of Olaudah* Equiano (1789), Frederick Douglass' *Narrative of the*

---

[6] As Lloyd observes, Agamben's theorization of the concentration camp as a permanent state of exception ignores the long colonial history of this camp.

Life of Frederick Douglass, an American Slave (1845), and William Wells Brown *Clotel; or, The President's Daughter: A Narrative of Slave Life in the United States* (1853). Today, partially thanks to the Federal Writer's Project's collection *Born in Slavery*, the corpus of texts narrated by people enslaved on US plantations is enormous.[7] These voices register how nineteenth-century capitalist society was experienced within the remits of the plantation as a central extractive commodity periphery. They thus testify to how this periphery operated as an engine of horrendous violence and constant personal, social, and material crisis.

As a long tradition of American Studies scholarship has made clear, people at the other end of semiperipheral, early nineteenth-century US society tell this story in a different, although not homogeneous, way. This era saw the publication of abolitionist writing such as Harriet Beecher Stowe's abolitionist bestseller *Uncle Tom's Cabin; or, Life Among the Lowly* (1852) which was critical of slavery, but that still often rehearsed profoundly racist tropes. These texts were challenged by anti-abolitionist, pro-slavery narratives such as Mary Henderson Eastman's *Aunt Phillis's Cabin* (1852), Caroline Lee Hentz's *The Planter's Northern Bride* (1854), and William Gilmore Simms' *The Sword and the Distaff* (1852), many of which engaged in direct polemic with Stove's novel. These texts were fiercely supportive of racialized slavery, they promoted the plantation as the main driver of the US economy and thus described it as a catalyst capable of elevating the US to a position of world hegemony. Many of these novels, disturbed by the enslaved Nat Turner's 1831 attempt to organize a violent revolution, also insist that the plantation and plantation slavery are necessary to protect white America from the violent urges of (enslaved) black people. In this way, the pro-slavery novel extends Underhill's thesis that non-white agency needs to be violently contained to safeguard extraction.

While deeply divided on the question of slavery and black agency, and what these mean for white society, both the abolitionist text and the

---

[7] See the Federal Writer's Project (https://www.loc.gov/collections/slave-narratives-from-the-federal-writers-project-1936-to-1938/). It can be added that Nicholas T. Rinehart (2019), building on Moretti's world-literature model, has proposed that 'enslaved testimony, like the novel, is a global phenomenon, even its own species of world literature' (11). Perceived as such, the slave narrative registers the human and ecological violence of the plantation from a clearly peripheral position within the world-system, and, from this position, this narrative describes the plantation as a locale that does not simply make violence possible, but that depends on violence for its function.

pro-slavery narrative register the ongoing transformation and depletion of land in oddly similar ways. While the settler colonial text is obsessed with the abject nature and enormous potential of unenclosed, Indigenous land, anti- and pro-slavery novels are both mainly concerned with land that has been cultivated by settler communities in the slave-owning South. An example is Tomas B. Thorpe's *The Master's House: A Tale of Southern Life* (1854), a novel that hovers uneasily between the abolitionist and pro-slavery novel. This narrative describes the life of the young, noble, and kind Graham Mildmay as he moves his house and his slaves from North Carolina to the more fertile Louisiana. Thus, a mostly unspoken but obvious premise of *The Master's House* is the fact that land can be exhausted by the extractive slave plantation. When first visiting Louisiana in search of the land that has, since the Indian Removal Act of 1832, been cleared of Indigenous people, Mildmay cannot 'suppress his enthusiasm at the richness of the vegetation he witnessed, and the easy manner with which they were made to produce an abundant crop, compared to the more sterile soil of North Carolina' (p. 42).

G. P. R. James' *The Old Dominion; or, The Southampton Massacre* (1856), an aggressively pro-slavery novel about the Nat Turner rebellion, is even more vocal on this issue:

> When we talk of a plantation, we think of a wide tract of country all smoothly laid out in maize, or tobacco, or cotton, or rice, and don't comprehend that perhaps two thirds of that plantation will be forest, either of the first or second growth. I must remark, too, that a good deal of country, especially on the seaboard, has gone back to forest, the earlier colonists having been like prodigals newly come into a fortune, and exhausted their lands with unvarying crops, principally of tobacco. Thus what was once, we have every reason to believe, very fertile soil, will only now bear pine or other trees of hardy habits. (p. 45)

Even if *The Old Dominion* does not theorize the relationship between extractivism, capitalism, slavery, and soil erosion, it does register, from a position of white privilege, local ecological breakdown. From the vantage of this privilege, this breakdown is regrettable and inconvenient, but in no way a reason to change direction or to worry about the long-term effects of extractive, slave-dependent agriculture.

Abolitionist writing also shows awareness of such breakdown, and it connects it to the extractive violence that occurs on the plantation more clearly than the pro-slavery, anti-Tom text. A telling example is Harriet Beecher Stove's novel *Dred: A Tale of the Great Dismal Swamp* (1856), written after Beecher Stove had read Frederick Douglass' fictional short story 'The Heroic Slave' (1852), where the escaped protagonist spends five years in a swamp to stay close to his wife. Inspired by Douglass' story and by the example set by Nat Turner (Turner's *Confessions* recorded in jail by the white attorney Thomas R. Gray (1831) are appended to the novel) and other black insurgents, the titular black character is not a docile Uncle Tom, but a rebellious, runaway slave named Dred. From the vantage of the Great Dismal Swamp that stretches across the border that separates Virginia and North Carolina, this character carries on a constant, low-intensity insurrection by rescuing other escaped slaves and by resisting slave society in other ways, all the while planning a revolutionary action of some sort. Thus, and unlike her first novel, *Dred* is a story about violent black resistance to extractive violence.

Scholarship in the novel has tended to centre the liminal territory of the swamp as this appears in the narrative. As Kathryn Benjamin Golden (2021) has shown, the Great Dismal Swamp was indeed a refuge for thousands of people who had escaped slavery. Vast, labyrinthine, and difficult for dogs to traverse, the swamp offered protection for a variety of societies: 'itinerant groups living along the swamp's peripheries', as well as 'intensely populated multigenerational communities, complete with their own methods and practices of self-government and community organization' (Golden 2021, p. 3). In this way, the swamp appears, in *Dred*, as an 'insurgent ecology' (Golden 2021, p. 8), a space out of which what the radical black minister Denmark Vesey called 'Natural Rights' can rise (Black 2022).[8]

However, *Dred* is equally a story about how slavery and the plantation regime exhaust fertile soil outside the swamp. The novel thus focuses on the slave-owning Gordon family which has, 'for two or three generations [...] lived in opulence' (p. 44). Following a period of 'gradual decay', these days are now over. The reasons why are clearly stated: 'Slave labor, of all others the most worthless and profitless, had exhausted the first

---

[8] Denmark Vesey was a black minister and community leader in early nineteenth century South Carolina. After winning the lottery, he bought his freedom, spoke fiercely about all people's 'natural rights', and allegedly began organizing a revolution that would allow enslaved people to escape to Haiti. He was condemned and executed for these plans in 1822 (Egerton 2004; Paquette and Egerton 2017).

vigor of the soil, and the proprietors gradually degenerated from those habits of energy which were called forth by the necessities of the first settlers' (p. 44). In this way, and as vaguely registered but not developed in *The Master's House* and *The Old Dominion*, the extractive violence of the plantation is seen as destroying not simply people, *but the land itself*. Similarly, the increasingly desperate attempts by the novel's incompetent plantation owners to manage the land and the enslaved people working on it only accelerate the suffering and the deterioration of the land. The violence wielded to bring enslaved people and the land to order is ultimately futile. Instead, the plantation itself is perceived as an engine of violence.

Even so, *Dred* is not a critique of capitalism as such. The North is described as doggedly and sustainably capitalist in the novel: 'If a man's field is covered with rock, he'll find some way to sell it, and make money out of it; and if they freeze up all winter, they sell the ice, and make money out of that. They just live by selling their disadvantages!' (p. 182). The novel's revolutionary message is also curbed when, at the end of the story, Dred's decision to attack the white community, and potentially shift society towards the ideals inherent in the insurgent swamp ecology he represents, is curtailed by the pure-hearted and loyal slave Milly. Thus, while the white core of antebellum American capitalist society was politically divided, the image of a successful black insurrection and the obliteration of the extractivist paradigm in favour of a fundamentally different social order was clearly difficult to insert even into an anti-slavery novel.

As the WReC has argued, 'the unfolding of combined and uneven development produces unevenness throughout the world capitalist system, and not merely across the divide represented by the international division of labour' (p. 57). Semiperipheries are processual spaces that testify to such unevenness, and that thus 'function not only hegemonically, transmitting value to core regions, but also counter-hegemonically, circulating new forms of solidarity and international consciousness across the global landscape of combined and uneven development' (WReC 2016, p. 549). This helps explain how pro-slavery writing, abolitionist texts, and autobiographies by enslaved people registered and narrated slavery in radically different ways, sometimes as a God-given and altruistic engine of white prosperity and black sanctuary, sometimes as an immoral, corrupting, and soil-depleting system, and sometimes as an impossibly violent and extractive machinery.

The world-literature model also assists in explaining how both aboli-
tion and pro-slavery writing register the ecological devastation that the
plantation produces, but avoids indicting extractive capitalism as the
engine of this devastation. Both types of text emerge out of white centres
of relative power; the slave-owning but agrarian South and the urban
and increasingly industrialized North. These represent two complementary, yet at the same time different and competing, evolutions of the US
semiperiphery. The question that plays out in these two different narratives is not simply if slavery should be allowed to persist, but if it is at the
agrarian commodity frontier, or within the wage-labour-driven, industrialized, and urban region, that capital should be primarily accumulated. This
charts two different paths forward for capitalism in the US, but neither
narrative imagines an alternative to extractive capitalism as such.[9]

EMERGENCY IN SETTLER CAPITALIST
WRITING AND THE PLANTATION TEXT

The texts discussed in this chapter emerge out of different stages of the
economic and social history of America, yet they speak in strangely similar
ways about capitalism, ecology, security, and violence. Underhill's *Newes
from America* registers settler capitalism via the violence performed on
Indigenous people and the land they inhabit and cultivate, but it simultaneously claims that this violence is both justified and necessary. Thus,
Newes from America insists that the emergency is located not in the killing
of Indigenous people or the destruction of the land they inhabit, but in
the possibility that these people and the land they live by may prevent
extraction. To symbolically 'spoil' and ruin the land is thus to make it
secure for extraction. In this way, the text registers the ecological and
human destruction that settler capitalism produces but envisions this as
a necessary response to the emergency that Indigenous people and land
create for the early settler capitalist enterprise.

The plantation text written by white authors from different sections
of the privileged semiperiphery also registers the ecological and human

---

[9] This is not to say that no such alternatives were imagined. An odd example is
George Fitzhugh's profoundly racist and fiercely pro-slavery treaty *Cannibals All!, or
Slaves Without Masters* (1857). This text is fundamentally critical of the impact that wage
labour and liberal capitalism has had on American people and society, and proposes slavery
as a utopian, colour-blind, socialist, Southern alternative.

violence that the plantation produces. In this way, the plantation text contributes both to the bourgeoning sense that extractive capitalism is endemically troubled by an emergency of its own making, and to the assumption that the solution to this emergency is the establishment of the security regime that a permanent state of exception enables. If extractive capitalism has engendered ecological and social erosion, the solution is not, in the pro-slavery text, the dismantling of the system as such, but the production of regimes capable of containing the constant eruption of crises. When this crisis is ecological, the solution is to move the location of the frontier; to find new land to extract. When the crisis is human and social, the solution is to build more effective systems to contain or eliminate rebellious elements. What must be secured is capitalism understood as the right to property, the enclosure, and the extractive processes that occur there. Even the anti-slavery texts that circulate an awareness of the social and ecological destruction produced by the plantation, and that gesture, like *Dred* does, to other types of sociality, fall short of promoting alternative relationships between (white) American society and the planet. In this way, both the pro-slavery and the abolitionist narratives promote an expansive capitalism as the only viable way forward for the burgeoning US nation-state. Thus, they fertilize the emergence of what Vandana Shiva (1993) has termed 'monocultures of the mind' (p. 5). As Shiva observes, such intellectual monocultures aid in the production of what Mark Fisher calls 'capitalist realism'. As described in the introduction, this realism generates a storyworld in which there is no alternative to capitalism even when this system is perceived to cause tremendous ruin and suffering. I will return to this concept in future chapters.

It should also be noted that the settler capitalist text and the plantation novel both register crises of sorts. Returning to Patel and Moore's observation that capitalism 'not only *has* frontiers; it exists only *through* frontiers, expanding from one place to the next' (19, italics in the original), it can be argued that capitalist crises typically occur in between the closing of one frontier and the opening of another. To simplify this process, the collapse of a particular frontier causes a dearth in the flow of resources from the frontier, while at the same time producing unemployment and a shortage of capital both among wage labourers and those accumulating the fruits of labour. This makes it necessary for capitalism to move on to new frontiers, to leave exhausted land behind and to find new fertile soil (Cheap Nature) where important cash crops can again be cultivated, or to find ways of reinventing a depleted frontier so that it can yield a new kind of commodity. Settler capitalist writing and the Plantation text both narrate this precise movement: how extractive capitalism must constantly move on to new frontiers as a response to the crisis caused by the exhaustion of the land. As I will argue in what follows, the American Climate Emergency Narrative is locked into the same logic, but comes out of a moment in time when the opportunity to relocate the frontier to keep sourcing nature cheaply is limited. Thus, this new narrative form gives voice and shape not simply to the imperative to extract and accumulate, but to the paralysing prospect that such accumulation may not be possible in the near future.

Works Cited

Acebo, Nathan P. 2021. 'Survivance Storytelling in Archaeology.' In *Routledge Handbook of the Archaeology of Indigenous-Colonial Interaction in the Americas*, edited by Lee M. Panich and Sara L. Gonzalez, 468–485. Abingdon: Routledge.

Agamben, Giorgio. 2005. *State of Exception*. Translated by Kevin Attell. State of Exception. Chicago: University of Chicago Press.

Aikens, Natalie, Amy Clukey, Amy K. King, and Isadora Wagner. 2019. 'South to the Plantationocene.' *ASAP Journal*. https://asapjournal.com/south-to-the-plantationocene-natalie-aikens-amy-clukey-amy-k-king-and-isadora-wagner/.

Anon. 1814. *The Charters and General Laws of the Colony and Province of Massachusetts Bay*. Boston: T. B. Wait and Co.

Arrighi, Giovanni. 1978. 'Towards a Theory of Capitalist Crisis.' *New Left Review* 111 (3): 3–24.

———. 1994. *The Long Twentieth Century: Money, Power, and the Origins of Our Times*. London: Verso.

Black, Christopher Allan. 2022. 'Vesey and Gordon's Righteous Insurrection: The Legacy of Denmark Vesey's Natural Rights Revolution in Harriet Beecher Stowe's *Dred: A Tale of the Great Dismal Swamp* (1856).' *Rocky Mountain Review of Language and Literature* 76 (2): 177–194.

Brandt, Stefan L. 2017. 'The American Revolution and Its Other: Indigenous Resistance Writing from William Apess to Sherman Alexie.' *AAA: Arbeiten aus Anglistik und Amerikanistik* 42 (1): 35–56.

Brown, William Wells. 1853. *Clotel; or, the President's Daughter: A Narrative of Slave Life in the United States*. 2016. Peterborough: Broadview Press.

Campbell, Tracy. 2005. *The Politics of Despair: Power and Resistance in the Tobacco Wars*. Lexington: University Press of Kentucky.

Carlson, Paul H. 1984. *Texas Woollybacks: The Range Sheep and Goat Industry*. College Station: Texas A&M University Press.

Cave, Alfred A. 1996. *The Pequot War*. Amherst: University of Massachusetts Press.

Davis, Heather, and Zoe Todd. 2017. 'On the Importance of a Date, or, Decolonizing the Anthropocene'. *ACME: An International Journal for Critical Geographies* 16 (4): 761–780.

Davis, Janae, Alex A Moulton, Levi Van Sant, and Brian Williams. 2019. 'Anthropocene, Capitalocene,… Plantationocene?: A Manifesto for Ecological Justice in an Age of Global Crises.' *Geography Compass* 13 (5): e12438.

Delahaye, Agnès. 2020. *Settling the Good Land: Governance and Promotion in John Winthrop's New England (1620–1650)*. Leiden: Brill.

Douglass, Frederick. 1845. *Narrative of the Life of Frederick Douglass, an American Slave*. 2018. Peterborough: Broadview Press.

———. 1852. 'The Heroic Slave. In *Autographs for Freedom*. Boston: John P. Jewett.

Drago, Harry Sinclair. 1970. *The Great Range Wars: Violence on the Grasslands*. New York: Dodd Mead.

Eastman, Mary H. 1852. *Aunt Phillis's Cabin; or, Southern Life as It Is*. Philadelphia: Lippincott, Grambo & Co.

Egerton, Douglas R. 2004. *He Shall Go Out Free: The Lives of Denmark Vesey*. Lanham: Rowman & Littlefield Publishers.

Equiano, Olaudah. 1789. *The Interesting Narrative of the Life of Olaudah Equiano*. 2001. Peterborough: Broadview Press.

Fitzhugh, George. 1857. *Cannibals All! Or, Slaves without Masters*. 1966. Cambridge, MA: Harvard University Press.

Foster, John Bellamy. 1999. 'Marx's Theory of Metabolic Rift: Classical Foundations for Environmental Sociology.' *American Journal of Sociology* 105 (2): 366–405.

———. 2002. *Ecology against Capitalism*. New York: New York University Press.

Golden, Kathryn Benjamin. 2021. '"Armed in the Great Swamp": Fear, Maroon Insurrection, and the Insurgent Ecology of the Great Dismal Swamp.' *The Journal of African American History* 106 (1): 1–26.

Gómez-Barris, Macarena. 2017. *The Extractive Zone*. Durham, NC: Duke University Press.

Grandjean, Katherine A. 2011. 'New World Tempests: Environment, Scarcity, and the Coming of the Pequot War.' *William and Mary Quarterly* 68 (1): 75–100.

Gray, Thomas R. 1831. *The Confessions of Nat Turner*. Baltimore: Lucas & Deaver.

Haraway, Donna. 2015. 'Anthropocene, Capitalocene, Plantationocene, Chthulucene: Making Kin.' *Environmental Humanities* 6 (1): 159–165.

———. 2016. *Staying with the Trouble: Making Kin in the Chthulucene*. Durham, NC: Duke University Press.

Hentz, Caroline Lee. 1854. *The Planter's Northern Bride*. Philadelphia: T. B. Peterson.

Holleman, Hannah. 2018. *Dust Bowls of Empire*. New Haven, CT: Yale University Press.

Holterhoff, Kate. 2019. 'Late Nineteenth-Century Adventure Fiction and the Anthropocene.' *Configurations* 27 (3): 271–300.

James, George Payne Rainsford. 1856. *The Old Dominion; or, the Southampton Massacre: A Novel*. New York: Harper & Brothers.

Kaplan, Jed O., Kristen M. Krumhardt, Erle C. Ellis, William F. Ruddiman, Carsten Lemmen, and Kees Klein Goldewijk. 2011. 'Holocene Carbon Emissions as a Result of Anthropogenic Land Cover Change.' *The Holocene* 21 (5): 775–791.

Koch, Alexander, Chris Brierley, Mark M. Maslin, and Simon L. Lewis. 2019. 'Earth System Impacts of the European Arrival and Great Dying in the Americas after 1492.' *Quaternary Science Reviews* 207: 13–36.

Lewis, Simon L., and Mark A. Maslin. 2015. 'Defining the Anthropocene.' *Nature* 519 (7542): 171–180.

Lloyd, David. 2012. 'Settler Colonialism and the State of Exception: The Example of Palestine/Israel.' *Settler Colonial Studies* 2 (1): 59–80.

Martin, John Frederick. 1991. *Profits in the Wilderness: Entrepreneurship and the Founding of New England Towns in the Seventeenth Century*. Chapel Hill: University of North Carolina Press Books.

Moore, Jason W. 2011. 'Ecology, Capital, and the Nature of Our Times: Accumulation & Crisis in the Capitalist World-Ecology.' *Journal of World-Systems Research* 17(1): 107–146.

———. 2015. *Capitalism in the Web of Life: Ecology and the Accumulation of Capital*. New York: Verso.

———. 2016. 'The Rise of Cheap Nature.' In *Anthropocene or Capitalocene? Nature, History, and the Crisis of Capitalism*, edited by Jason W. Moore, 78–115. Oakland, CA: PM Press.

Nevle, Richard J., and Dennis K. Bird. 2008. 'Effects of Syn-Pandemic Fire Reduction and Reforestation in the Tropical Americas on Atmospheric CO_2 During European Conquest.' *Palaeogeography, Palaeoclimatology, Palaeoecology* 264 (1–2): 25–38.

Paquette, Robert L., and Douglas R. Egerton. 2017. *The Denmark Vesey Affair: A Documentary History*. Gainesville, FL: University Press of Florida.

Parker, Geoffrey. 2013. *Global Crisis: War, Climate Change and Catastrophe in the Seventeenth Century*. New Haven: Yale University Press.

Patel, Raj, and Jason W Moore. 2018. *A History of the World in Seven Cheap Things: A Guide to Capitalism, Nature, and the Future of the Planet*. London: Verso.

Rinehart, Nicholas T. 2019. *Narrative Events: Slavery, Testimony, and Temporality in the Afro-Atlantic World*. PhD thesis. Harvard University.

Shiva, Vandana. 1993. *Monocultures of the Mind: Perspectives on Biodiversity and Biotechnology*. London: Zed Books.

Shogan, Robert. 2004. *The Battle of Blair Mountain: The Story of America's Largest Labor Uprising*. New York: Basic Books.

Simms, William Gilmore. 1852. *The Sword and the Distaff; or "Fair, Fat and Forty,"*: *A Story of the South, at the Close of the Revolution*. Philadelphia: Lippincott, Grambo, & Co.

Slotkin, Richard. 1973. *Regeneration through Violence: The Mythology of the American Frontier, 1600–1860*. Norman: University of Oklahoma Press.

Starr, Timothy. 2012. *Railroad Wars of New York State*. Charleston, CS: The History Press.

Steffen, Will, Paul J. Crutzen, and John R. McNeill. 2016. 'The Anthropocene: Are Humans Now Overwhelming the Great Forces of Nature?' In *The New World History*, edited by Ross E. Dunn, Laura J. Mitchell and Kerry Ward, 440–459. Berkeley, CA: University of California Press.

Stowe, Harriet Beecher. 1852. *Uncle Tom's Cabin: Or, Life among the Lowly*. 2009. Cambridge, MA: Harvard University Press.

———. 1856. *Dred: A Tale of the Great Dismal Swamp*. Vol. 1. Boston: Phillips, Samson and Co.

Thorpe, Tomas B. 1854. *The Master's House: A Tale of Southern Life*. New York: T. L. McElrath & Co.

Tilley, Lisa. 2020. '"A Strange Industrial Order" Indonesia's Racialized Plantation Ecologies and Anticolonial Estate Worker Rebellions.' *History of the Present* 10 (1): 67–83.

Underhill, John. 1638. *Newes from America; or, a New and Experimentall Discoverie of New England*, edited by Paul Royster. 2007. Lincoln: Libraries at University of Nebraska-Lincoln. https://commons.und.edu/cgi/viewcontent.cgi?article=1048&context=settler-literature.

Wagner, D. R., and J. Dempsey. 2004. *Mystic Fiasco How the Indians Won the Pequot War*. Scituate, MA: Digital Scanning.

Williams, Kyle. 2020. 'Roosevelt's Populism: The Kansas Oil War of 1905 and the Making of Corporate Capitalism.' *The Journal of the Gilded Age and Progressive Era* 19 (1): 96–121.

Winegard, Timothy C. 2016. *The First World Oil War*. Toronto: University of Toronto Press.

Wolfe, Patrick. 2006. 'Settler Colonialism and the Elimination of the Native.' *Journal of Genocide Research* 8 (4): 387–409.

Wolford, Wendy. 2021. 'The Plantationocene: A Lusotropical Contribution to the Theory.' *Annals of the American Association of Geographers* 111 (6): 1622–1639.

Warwick Research Collective (WReC). 2016. 'WReC's Reply.' *Comparative Literature Studies* 53 (3): 535–550. https://doi.org/10.5325/complitst udies.53.3.0535. Accessed 2023/04/06.

Yusoff, Kathryn. 2018. *A Billion Black Anthropocenes or None.* Minneapolis: University of Minnesota Press.

Zalasiewicz, Jan, Colin N. Waters, Mark Williams, Anthony D. Barnosky, Alejandro Cearreta, Paul Crutzen, Erle Ellis, Michael A. Ellis, Ian J. Fairchild, and Jacques Grinevald. 2015. 'When Did the Anthropocene Begin? A Mid-Twentieth Century Boundary Level Is Stratigraphically Optimal.' *Quaternary International* 383: 196–203.

CHAPTER 3

Fossil Fictions

FOSSIL CAPITALISM AND FOSSIL FICTION

The intensification of the slave economy in the early American republic coincided with the global rise of what Andreas Malm (2016) calls 'fossil capital'. The turn to fossil energy vastly increased capitalism's ability to colonize and settle land and to extract nature cheaply from it. This, in turn, transformed, expanded, and accelerated capitalism as such. As Malm argues, when steam power was introduced, the energy it supplied was not necessarily more abundant or cheaper than water or wind, but steam did provide the opportunity to insert labour into the production process in novel and, for capitalism, ultimately very profitable ways. Freed from the energy limitations that wind, water, and sun had previously imposed, the capitalist world-system was able to spread its global presence more widely and effectively. It is no coincidence that the latter half of the nineteenth century, when steam had become the main accelerant of the world-system, has been termed the Age of Empire (Hobsbawm 1987). During this era, the colonial/capitalist system was spread to virtually all parts of the world until, shortly after the turn of the century, some 84 per cent of the world's land surface had been formally colonized by European nation-states (Atkin and Biddiss 2008, p. 177).

The turn to fossil energy and the consequent establishment of new commodity frontiers created new possibilities, comforts, and privileges for

© The Author(s) 2024
J. Höglund, *The American Climate Emergency Narrative*,
New Comparisons in World Literature,
https://doi.org/10.1007/978-3-031-60645-8_3

the US settler community, but it also produced new insecurities and emergencies, and thus also of new ways of managing these. Coal demanded a great deal of labour and work at the coal commodity frontier, or in the many coal-fuelled industries that grew up around urban centres, was often very dangerous. Meanwhile, the lack of, or reluctance to obey, labour laws often meant that workers were cheated out of their fair wages. From the perspective of capital, the emergence of a vast, interracial, unionized, and frequently militant body of labour was a considerable problem to profitable extraction and it was often addressed, as the chapter shows, through (para-)military campaigns. In addition to this, the rise of fossil capitalism nurtured already existing tensions between the dominant nation-states within the expanding world-system. With the emergence of the fossil-energized military-industrial complex, the production of instruments of war, and war itself, became increasingly important to capitalism. This complex facilitated new and uniquely deadly and mobile types of weapons such as the ironclad warship and, after the turn to oil, the tank, and the aeroplane. In view of this development, it is not surprising that the first half of the twentieth century saw two uniquely deadly world(-system) wars.

These insecurities and capitalism's attempts to manage them were registered by US literature and early films written and produced from what, by the early twentieth century, had become the new core of the world-system. This chapter thus focuses on films and literature from the US first during its transition from an ambitious semiperipheral nation into a core nation, and then as the new core of the world-system. As discussed, this is a historical development profoundly marked by the tremendous acceleration of capitalism's ability to generate capital and privilege, but also by new and increasingly ecocidal extractive systems and by the two world wars that demonstrate the enormous destructive potential of fossil capital. The chapter discusses in particular how the fiction under scrutiny registers the systematic social and ecological violence that characterizes these specific stages, and how it narrates this violence as an inevitable response to emergency and crisis. This body of texts can be termed fossil emergency fiction, a concept that references Andreas Malm's (2017) 'fossil fuel fiction' described by him as 'fiction dealing with fossil fuels and imagining some disaster linked to them' (126). Malm's term brings the systemic processes that have caused global warming to the fore, and it thus functions as a important corrective. Fossil emergency fiction does essentially the same, but unlike the stories that Malm's work centres on, my focus is exclusively on fiction from the core and how it narrates the various emergencies that coal and oil are perceived to both create and resolve. Literature and film produced from this vantage not only register crises linked to fossil fuel, they also suggest fossil resolutions to these crises.

In what follows, the chapter first turns to the establishment of the coal commodity frontier and the intense human and ecological violence enacted there. This is followed by a section that looks specifically at the (military) potential, insecurities, and violence afforded by the turn to oil. While the rise of the US within the world-system has been registered in a plethora of ways in fiction, and from very different positions within this world-system, these particular narratives clearly build on the legacy left by the early settler capitalist text and by the white plantation narrative. Based on the same extractive logic as these earlier types, and taking place in similar capitalist storyworlds, the texts analysed in this section also seek to sanitize the increasingly profound ecological and human violence utilized as responses to the crises that drive the story. In this way, they constitute important stages in the emergence of the American Climate Emergency Narrative.

COAL FRONTIERS AND COAL INSECURITIES

The first industrial revolution, accelerated by coal, iron, and cheap labour, arrived relatively late in the US. When it did, it was a development partially provoked by the military and economic emergencies generated by the War of 1812. Following the rise of Napoleon, Britain sought to limit trade between France and the US while at the same time protecting its hold over Canada and resisting further expansion of the US into the American West. In 1814, as part of this effort, the British marched to Washington where they burned several government buildings; still the only occasion when US borders have been breached by a foreign army. With its access to European industries and markets compromised, the US decided to fast-track the construction of its own industrial and economic infrastructure. This project was termed *the American System* and included tariffs on imports, the establishment of national banks, and the construction of railroads and steam-powered shipping routes that connected producers of raw materials with industries, consumers, and the international market.

In 1832, Secretary of State Henry Clay—the main architect of the American System—celebrated the material wealth created by the restructuring of US banking, infrastructure, and industry he had helped accomplish. From the vantage of the emerging core, he produced the following image of a transforming US:

> whole villages [are] springing up, as it were, by enchantment; our exports and imports increased and increasing; our tonnage, foreign and coastwise, swelling and fully occupied; the rivers of our interior animated by the perpetual thunder and lightning of countless steamboats; the currency sound and abundant; the public debt of two wars nearly redeemed; and, to crown all, the public treasury overflowing. (Clay 1857, p. 440)

In this description, coal and capital perform a kind of striking (white) magic that transforms and modernizes the agrarian state. Out of the pastoral and debt-ridden America of the early republic, an urban utopia rises, literally out of the ground. This is early carbon modernity as progress and privilege; as the engine of capitalist prosperity and of national financial and military security.

While coal transformed life for the emerging white core of the American republic, it also produced a new set of insecurities that this white settler community had to consider. Coal does not lend itself to the corporeal metaphor that describes fossil fuels as the lifeblood of American capitalism as easily as oil. Yet, as Peter A. Shulman observes in *Coal and Empire: The Birth of Energy Security in Industrial America* (2015), it was as important to nineteenth-century US industry, commerce and war-making as oil was during the late twentieth century. As Shulman argues, '[w]hen seen from the perspective of coal, the great process of industrialization and the emergence of the United States as a global power unfolded at the same time as intertwined processes' (p. 6). As the energy that fuelled industry, that kept merchant vessels, trains, and navy ships running, it was coal that made it possible for the US to take an increasingly central position within the world-system.

This put coal at the centre of discourses and systems of securitization. To keep industry and trade going, and to enforce the 1823 Monroe Doctrine that aimed to keep European powers out of the western hemisphere, extraction at the coal commodity frontier had to be made secure. From the early nineteenth century and onwards, this was a major concern for the federal government, local industries, state and private militias, lawmakers, and writers within the evolving US semiperiphery. In Washington, 'the American approach to energy' was thus shaped by 'politicians and policy makers' and by 'naval administrators and officers, who played central roles in articulating the significance of coal for the navy' (Shulman 2015, p. 8). At moments of national or international tension, the flow of coal from coal mines to the navy became a crucial imperative. In particular, a steady supply of coal was essential during the Civil War and in connection with the attempt at formal empire building that followed in the wake of the Spanish–American war in 1898.

This, in turn, put pressure on local contractors tasked with delivering coal to industry and the military at a low cost. Because the mining of coal was extremely hard work, this meant securing access to (cheap) labour. In the antebellum South, enslaved people were frequently utilized to work in the coal mines (Lewis 1987; Adams 2004). After the war, formerly enslaved people remained a crucial labour resource.[1] For many, the first step away from the plantation led to the mines and industries that had grown up close to these mines. In parts of the South, and increasingly in the North, coal mining was also performed by poor white communities and by newly arrived, unskilled European immigrants.

In writing from the semiperipheral commodity frontiers where coal was extracted, this development is, not surprisingly, described as disorienting and violent. In the opening chapters of Booker T. Washington's autobiography *Up From Slavery* (1901), the author describes how he worked as a child labourer in the coal mines of West Virginia. In this text, the coal mine is narrated as a confusing and labyrinthine space:

> I do not believe that one ever experiences anywhere else such darkness as he does in a coal-mine. The mine was divided into a large number of different "rooms" or departments, and, as I never was able to learn the location of all these "rooms," I many times found myself lost in the mine. To add to the horror of being lost, sometimes my light would go out, and

---

[1] The need for cheap (coal) energy and cheap labour thus became entangled with the fundamentally racist effort to build a post-War, white America. The 'colonization' project that sought to exile formerly enslaved people to Africa, South America, and other parts of the world was used to enable coal mining in these locales. One striking example was the colonization of Chiriquí region of Colombia (today Panama), the purpose of which was to send black people to a coal-dense valley in the region and to use these people as cheap and precarious labour (see Page 2011).

then, if I did not happen to have a match, I would wander about in the
darkness until by chance I found some one to give me a light. (p. 38)

In addition to this, it is also a location that promotes both physical and
intellectual death:

There was always the danger of being blown to pieces by a premature
explosion of powder, or of being crushed by falling slate. Accidents from
one or the other of these causes were frequently occurring, and this kept
me in constant fear. Many children of the tenderest years were compelled
then, as is now true I fear, in most coal-mining districts, to spend a
large part of their lives in these coal-mines, with little opportunity to
get an education; and, what is worse, I have often noted that, as a rule,
young boys who begin life in a coal-mine are often physically and mentally
dwarfed. They soon lose ambition to do anything else than to continue as
a coal-miner. (pp. 38–39)

These passages can be considered a straightforward attempt by Washington to describe his encounter with the coal commodity frontier, but it
is also possible to read it as a sequence that registers, from the perspective of the (semi)periphery, a more general sense of being lost within
a new economic and ecological space. Washington has managed to exit
the agrarian slave plantation environment only to be introduced into
an underground world of darkness, insecurity, and confusion. In other
words, the passage shows Washington folded into the world-system as
cheap labour but also as 'Cheap Nature'; a semi-human resource to be
extracted. It should be added that it was Washington's cheaply paid
extractive work, but also his position as an extractable resource, that made
Clay's vision of an urban American utopia, magically rising from the land,
possible.

In this way, and considering both Clay's and Washington's descriptions
of early fossil capitalism, texts from the US sometimes register the abject
violence of the coal economy, but they also at times elide it or displace
it as a kind of magic. Stephanie LeMenager (2014) has observed that
when oil became 'an expressive form', oil itself was often 'hidden [...] in
plain sight' (p. 66). In the same way, American late nineteenth- and early
twentieth-century fiction is about coal also when it does not make this
energy form its main concern. American writing from the period thus
typically narrates coal energies from the same core perspective as that
employed by Clay, focusing on the new type of urban modernity that

coal made possible. In other words, coal is present in stories that are not about coal extraction, but about the wealth, privileges, romances, and comforts it produces for certain strata of American society. Such writing registers the enormous wealth that early carbon modernity generated, but casts it, like Clay does, as a strange, enchanted unfolding of the American republic.[2]

However, there is also an extensive body of writing from the turn-of-the-century US semiperiphery that makes coal extraction and coal burning clearly visible and that directly registers the violence used to secure cheap labour. These narratives make visible the securitization strategies designed to contain land and people, and, as in the plantation text, they make visible some of the ecological havoc that coal mining generated (and generates). Yet, like the plantation text, these fictions still envision extraction as inevitable and progressive. A case in point is the fiction of John Fox Jr., one of the first American authors to write what today is considered a bestseller. His first commercially successful novel was *The Little Shepherd of Kingdom Come* (1903). This story about Kentucky during the Civil War was adapted into films in 1920, 1928, and 1961 before its rehearsal of racist tropes became too politically problematic for Hollywood to adapt. These tropes are also present in his bestselling *The Trail of the Lonesome Pine* (1908), a novel similarly adapted into several films by the early Hollywood industry. This text, like much of Fox's other writing, is set on the coal commodity frontier.

Fox had a stake in this frontier. As Darlene Wilson (1995) has observed, he and his half-brother James 'were involved in developing the modern southern coal industry, particularly in the bituminous coal fields of Old Virginia, northeast Tennessee, and eastern Kentucky' (p. 6). Thus, Fox writes in his own interests and from a space of capitalist opportunity and privilege. Fox's writing is a part of the recasting of America from a rural nation to an urban carbon society, but its focus is very much on the coal commodity frontier itself, narrated as a pioneer frontier landscape. As Henry D. Shapiro (1986) and Rodger Cunningham (1990) have noted,

---

[2] It should be added that the nineteenth century novel as a commodity and a past time testifies to the advent of coal modernity. It is the coal-driven industries (printing presses), and the leisure enjoyed by the growing, white middle class that made the novel into the dominant art form of the core and privileged parts of the semiperiphery at this time.

Fox's stories rehearse the established colonial saga where white, masculine figures enter a both primitive and feminized landscape inhabited by uneducated and belligerent people engaged in brutal tribal violence.

The Trail of the Lonesome Pine exemplifies this trend through its description of the hero protagonist, coal prospector John Hale, as a 'by instinct, inheritance, blood and tradition—pioneer' (p. 40). Thus, Fox casts the coal prospector as a next-generation settler who enters a territory where the frontier that Fredrick Turner (1893) celebrated as an important but vanished driver of American societal evolution can be rediscovered. This frontier space is inhabited by poor white, possibly racially mixed, people who stand in for, but also effectively erase, the Cherokee, Yuchi, and Shawnee people who used to live in the mountains. Hale has arrived on this landscape because he is in search of coal, and he soon locates an especially rich vein on the land of a local family named Tolliver: 'that coal, cannel, rich as oil, above water, five feet in thickness, easy to mine, with a solid roof and perhaps self-drainage, if he could judge from the dip of the vein' (p. 35). Unlike the Tollivers, Hale understands, and is in fact an agent of, coal's relationship to the world-system. Looking at the coal, he knows that there is 'a market everywhere—England, Spain, Italy, Brazil' (p. 35). Hale does feel a pang of guilt for taking advantage of this knowledge: 'if he would take the old man's land for a song—it was because others of his kind would do the same!' (p. 36). Indeed, other prospectors are closing in: 'The English were buying lands right and left at the gap sixty miles southwest. Two companies had purchased most of the town-site where he was—HIS town-site—and were going to pool their holdings and form an improvement company' (p. 127). When Hale and other prospectors have finished buying up the land and started to extract coal, people and society rapidly transform. June, the daughter of the primitive Tolliver family, undergoes a '[m]agic transformation' (p. 227) much like the city in Clay's description, and in the previously impoverished local community money is suddenly as 'plentiful as grains of sand' (p. 234). Hale finds himself 'on the way to ridiculous opulence and, when spring came, he had the world in a sling and, if he wished, he could toss it playfully at the sun and have it drop back into his hand again' (p. 233).

However, in other passages, it is clear that the coal frontier also produces considerable social and ecological havoc. Coming back to the house and land where she grew up, June Tolliver notes how the harsh yet pristine mountain has been transformed:

[T]he willows bent in the same wistful way toward their shadows in the little stream, but its crystal depths were there no longer—floating sawdust whirled in eddies on the surface and the water was black as soot. Here and there the white belly of a fish lay upturned to the sun, for the cruel, deadly work of civilization had already begun. Farther up the creek was a buzzing monster that, creaking and snorting, sent a flashing disk, rimmed with sharp teeth, biting a savage way through a log, that screamed with pain as the brutal thing tore through its vitals, and gave up its life each time with a ghost-like cry of agony. Farther on little houses were being built of fresh boards, and farther on the water of the creek got blacker still. (pp. 201–202)

While this passage is short, it is also notably visceral in its description of the destruction of the local ecosystem. The passage focuses on the violent ecological transformation of a small section of the mountain, but this transformation is, of course, connected to what is going on all around the world. Because Appalachian coal has already been linked to a world market ('England, Spain, Italy, Brazil' [p. 35]), the ecological emergency briefly described is clearly tied to the world-system. Like the plantation text, then, this passage registers extraction as detrimental to the land. It is a fleeting yet striking sequence that can be read as an eruption of what Fredrick Jameson (1981) has termed the political unconscious that fissures even the capitalist, pro-extraction novel. In other words, the passage acknowledges that extractive capitalism, here termed the 'work of civilization', erodes ecology. At the same time, the novel in its entirety never suggests that this civilization is optional. Deadly as it is, its evolution is described as progressive and inevitable.

The extractive practices performed on the coal commodity frontier also provoke social violence in the novel. Workers at a quickly erected brick plant call a sudden strike: 'armed with sticks, knives, clubs and pistols, they took a triumphant march through town' (p. 93). In response to this sudden crisis, Hale creates an improvised police force. This is to be the instrument by which the strikers and other unruly elements are reinserted into extraction. Hale engages friends from the valley to help in this work. They are described as 'Bluegrass Kentuckians ... of pioneer, Indian-fighting blood' (p. 88). Thus, the novel connects the improvised police force with former soldier settlers such as Underhill, responding to the insecurities created by Indigenous people. At the same time, Hale and his fellow officers look to the Ku Klux Klan for inspiration. This organization, 'they all knew [...] had been originally composed of gentle-men' (p. 95). In this way, the novel also connects the budding police to the post-Civil War effort to undo the civil liberties gained by formerly enslaved plantation workers.

Alex S. Vitale (2017) has argued that the primary function of police forces in (Anglo) society has been to 'maintain political control and help produce a new economic order of industrial capitalism' (p. 36). Hale may be the just and vigilant hero of the novel, but it is still clear that his rein-vention of policing on the Appalachian coal commodity frontier is geared towards the imminent need to create a social, judicial, and paramilitary order that allows extractivist capitalism to do its work.[3] In other words, Hale's police force is another strategy designed to manage some of the insecurities that extraction and the law of Cheap Nature create. Hale (and the novel in its entirety) insists that this adaptation is fundamentally legal, but the genealogy of the force, its roots in settler capitalist violence and the Ku Klux Klan, reveals how this effort is underwritten by a funda-mentally racist system of indiscriminate violence invented as an attempt to manage Indigenous people during settlement, and black agency after the Civil War. As such, this force introduces not the law as much as a permanent and extra-legal state of exception.

In the novel, the police force does indeed resolve the emergency that insurgent labour poses. Extraction continues until all the coal has been dug from the land and until those resisting extraction have died violent deaths or given up. In the final passage, Hale unites with a now thor-oughly civilized June, cementing the notion that the coal commodity frontier is also capable of revitalizing people at the site of extraction. In this way, the novel exemplifies how writing from the privileged US semiperiphery makes use of tropes inherited from both the settler colo-nial text and the plantation novel. Like these earlier narratives, *The Trail of the Lonesome Pine* registers how extraction produces disturbing ecolog-ical and social devastation, and it leverages (displaced) indigeneity and unwilling labour as agents of *crises*. At the same time, and again like these prior literary forms, it describes extractive capitalism as essential and inevitable, and poses (para) military violence as the only way to manage

---

[3] In many ways, Hale's creation of a strong police presence capable of defending his own investment in coal mirrors the creation of the privatized Pennsylvania Coal and Iron Police, formed in 1865 and used, until 1931, to protect the interest of capitalist entrepreneurs (see Sadler 2009).

crises. These are elements that have long been rehearsed by the settler capitalist and plantation text, and they are also clearly carried over into *The Trail of the Lonesome Pine* as another forerunner of the American Climate Emergency Narrative.

THE PETROWAR OIL ENCOUNTER

In *World Literature and Ecology*, Michael Niblett (2020) describes how the limits of the coal economy, the depression of the 1930s, and the (global) Dust Bowl, forced capital to reorganize labour, land, and extraction. The 'lifeblood', Niblett observes, 'of this new phase of capitalism and the specific forms of mass production, consumption, and transport it involved' (p. 206) was oil. The British Empire, already contracting in the wake of WWI and beset by (coal miner) strikes at home, was unable to effectively adapt to this new phase so that, as Niblett puts it, the 'new regime of accumulation' was 'dominated by the United States' (p. 206). This signals the moment when the US finally replaced a contracting British Empire as the hegemon of the world-system. From this moment on, a great deal of American writing can be said to emerge out of the self-confident core of this world-system, even if the US is still made up of regions that remain semiperipheral to this new core.

At this stage, oil had begun to accelerate all aspects of the world-system, including its ability to burn massive quantities of fossil fuel and, as a consequence, release greenhouse gases into the biosphere. In the interwar years, the US was able to source most of its oil from the semiperipheries that existed within its own borders, but after North American oil production peaked in 1971, the US grew increasingly dependent on 'foreign oil' (Painter, 2014). Before long, the need to acquire oil as quickly and cheaply as possible demanded a new type of energy geopolitics, one capable of opening up Latin America, Africa, and the Middle East to oil companies located in the Global North. This, in turn, created a new set of insecurities, ecological emergencies and crises, as well as a revised set of (military) strategies and technologies by which these were managed.

In other words, oil transformed and energized how extractive capitalism operated across the planet, while at the same time revolutionizing the tools used to secure extraction. Specifically, oil was at the centre of the military's invention of new ways to access (enemy) space, new types of weapons, and new ways of combating not just the enemy but the very geography (forests, jungles, cities) in which the enemy was located, and, in the process, it produced new ways of perceiving the planet and its people. These technologies were made possible by oil, but they were also tools through which oil could be secured. The increasing centrality of the dispersed oil commodity frontier for American capitalism and the evolution of war technologies that oil enabled (and enables), make it possible to talk about a new kind of warfare. From this moment on, many wars are *Oil Wars* or *Petrowars*. This is a war fought using a war machine accelerated by oil, and it is also a geographically dispersed war *for* oil. In this way, Petrowar functions as a kind of circular economy of energy, capital, securitization, and death where the ultimate goal of Petrowar is to secure access to the energy form that fuels it and the society it serves. Most importantly for this study, the advent of Petrowar greatly enhanced the capacity of the war machine to kill people, control labour, level architecture, and destroy the land people inhabit.

In *The Shock of the Anthropocene*, Christophe Bonneuil and Jean-Baptiste Fressoz (2016) use the concept of the Thanatocene to highlight how central a part the military has played in the history of socio-ecological breakdown. They note that the increase in the human death toll during the wars fought by the US during the twentieth century was accompanied by an increase in the burning of fossil fuels (and thus also in the amount of CO_2 released into the atmosphere). During WWII, the US army, under the command of General Patton, consumed 1 gallon or 3.7 litres of petrol per person per day. During the Vietnam War, this figure had increased to 9 gallons or 33.2 litres, and during the 2003 invasion and occupation of Iraq, a total of 15 gallons or 55.5 litres of petrol per person per day were used. As Bonneuil and Fressoz also observe, an Abrams tank uses 40 litres of fuel per 10 kilometres while the B-52 bomber uses 12,000 litres of jet fuel per hour. In the present moment, the US Department of Defense, despite its endeavour to limit its use of oil, remains the world's single largest consumer of hydrocarbons (Crawford 2022, pp. 7–8), and the CO_2 it releases into the atmosphere signifies an ancillary planetary violence that adds to global warming (Bonneuil and Fressoz 2016, pp. 123–124).

Bonneuil and Fressoz also note how, during the twentieth century, Petrowar technologies made it possible to effectively make war on the planet itself. Petro-weapons such as napalm were invented to make it possible to target not just individual people or even groups of people, but entire areas where people might be sheltering. Napalm was thus used to destroy fields, forests, waterways, and other ecosystems vital to the long-term survival of the enemy. This was done with increasing intensity during both world wars until, during the Vietnam War, Bonneuil and Fressoz observe, an 'estimated 85 per cent of the ammunitions used by the US Army were targeted not at the enemy but at the environment sheltering them: forests, fields, cattle, water reserves, roads and dikes' (p. 127). During the same war, the US also employed the herbicide Agent Orange to de-leaf and destroy vast forests. Indeed, if the Vietnam War was an attempt to securitize Asia from communist influence, and thus ensure that other Asian countries would remain cogs in the capitalist world-system, this goal was accomplished through large-scale violence done to Vietnamese forests, waterways, animals, and fields. To securitize the extractive and capitalist world order that Captain John Underhill paved the way for, as discussed in Chapter 2, Underhill's strategy of 'burning and spoyling the Countrey' (1638, p. 14) was applied on a global scale.

PETROWAR FICTION

As the energy that ushered the US into an era of world-system dominance, oil has had an enormous impact on world culture. In a review from 1992 (a time when the US and allies had recently ceased the hostilities in the Persian Gulf that have been named the First Gulf War), Amitav Ghosh coined the influential concept of Petrofiction to describe texts that investigate 'the Oil Encounter' (p. 29). In this review, Ghosh argues that, despite the centrality of oil to America and the world, the Oil Encounter has produced 'scarcely a single work of note' (p. 29), and he speculates that this may be because to Americans 'oil smells bad' (p. 30). Problematizing both these claims, Stephanie LeMenager argues in *Loving Oil, Petroleum Culture in the American Century* (2014) that 'the story of petroleum has come to play a foundational role in the American imagination' (p. 5). In her study, she pays particular attention to the 'charisma of energy' and notes that while '[c]oal fictions emphasize labour struggles, the potential power of the strike, and solidification of a working class', the petrotext narrates the 'materialization of the liberal tradition in middle-class self-possession' (p. 5). By doing so, petroleum is encountered everywhere: in the apartment store, the automobile, the cinema, and the jet aeroplane. Literature set in such places can also be called Petrofiction, but unlike the type of Petrofiction that Ghosh can find little trace of, it is written at the core rather than at oil's peripheral or semiperipheral commodity frontier.

Because American writing is saturated by oil it would be possible to investigate a plethora of very different Petrotexts from the core. Again, some of these texts would register the pleasures that oil makes possible, while others would narrate the US exploitation of oil peripheries in the Global South as masculine adventure. However, the type of narrative that most interestingly continues the kind of ideological work performed by the settler colonial text, the plantation novel, and the coal frontier text, and that thus paves the way for the American Climate Emergency Narrative, is the *Petrowar* text. Oil may be present in virtually all post-WWII American fiction in some way, but oil as well as the attempt to secure oil extraction, is arguably most directly encountered on the battlefield. Indeed, the meeting with the bomber aeroplane and the tank during WWI were direct and brutal Oil Encounters. The German WWII 'Blitzkrieg' that used highly mobile motorized infantry in combination with dive bombers was completely organized around oil's capacity to accelerate war. The attack on Pearl Harbor was provoked by the US' attempt to limit Japan's access to oil in the Dutch East Indies (Feis 1950; Stinnett 2001), and the aerial attack itself was also an Oil Encounter. Similarly, Dresden during the Allied firebombing, London during the Blitz, the Vietnamese villages that were showered with napalm during the Vietnam War, and the bombing of parts of Iraq in 1991–1992 and again in 2003, were sites of the Petrowar Oil Encounter.

American fiction from the core has been enormously fascinated with Petrowar and there are an extensive number of novels, films, songs, artworks, and games that narrate it. These texts register the violent effect that Petrowar has on earth-bound soldiers and civilian life, but they are equally interested in the oil-accelerated machines of war: the submarine, the battleship, the tank, the jeep, and, above all, the combat aeroplane. In such texts, the oil encounter is the meeting with the petroleum-fuelled machines of modern warfare, with the enormous destructive power they bring, and with the new visual vantages they introduce: the aerial perspective afforded by the fighter plane, the periscopic view of the submarine, the gun-aim of the tank. Like pornographic films or slasher horror, the plot is not necessarily the point of this type of text. Rather, these films' main subject is the demonstration of the oil-accelerated machines of war. A central moment of the Petrowar text is when the Petrowar machine embraces the soldier, allowing him (or, on the rare occasion, her) to wield the astoundingly destructive potential of oil. As part of the combat aeroplane or the submarine, the soldier is afforded new mobilities and new ways of performing violence. In this way, the effort to securitize extractive capitalism on a planetary scale is narrated as a merger between a militarized agent of capitalism and the Petrowar machine.

One of the most striking and widely disseminated Petrowar texts is unarguably *Top Gun* (1986), directed by Tony Scott. To be able to shoot the extensive aerial action sequences, the film made use of the increasingly symbiotic relationship between the entertainment industry and different military branches of the US Department of Defense. This relationship has been termed by James Der Derian (2001) the Military-Industrial-Media-Entertainment network (MIME-Net) while Mackenzie Wark (2007) and Tim Lenoir and Henry Lowood (2005) call it the Military Entertainment Complex.[4] As briefly mentioned in the introduction, this Complex is essentially a circular economy where the narrative is doctored to stimulate the industrial production of military hardware, recruitment of soldiers, and actual military engagement. This hardware, these soldiers, and these engagements then produce new narratives that energize another turn of the cycle. In this way, the capitalist core directly intervenes in the production of narrative.

Top Gun can thus be described as a text that erupts directly out of the material and political core of the hegemonic world-system. In return for expensive aircrafts, the film conveys an image of the US Petrowar machine as enormously powerful, effective, and precise. The protagonist of the film is not really Tom Cruise's hell-raising pilot, but the petro-accelerated war machine it features. *Top Gun* thus opens with images of fighter jets being serviced and sent off into an early dawn. Smoke billows around the jets as the sun rises and the morning sun is then eclipsed by the roaring bright fire coming out of the twin tail engines. Steel wires in the start and landing strip reveal that these jets are taking off from an aircraft carrier. People clad in helmets and overalls move around the jets, dragging hoses with aircraft fuel, tuning wheels, and celebrating a successful landing with impromptu dance moves. This is a film about accelerated American Petrowar on a planetary scale. Set on a mobile war platform traversing the Indian Ocean, within reach of Middle Eastern oil fields, fighter jets

---

[4] See also Tim Lenoir (2000), David Robb (2004), Nick Turse (2008), and Johan Höglund (2008, 2012).

incessantly take off and land—ever vigilant and potent—fuelled by the oil economy they have (in fact) been built to protect (Fig. 3.1).

Fig. 3.1 Scenes from the opening of *Top Gun* (1986) showing fighter jets taking off, being refuelled, and landing

The assistance that the Military Entertainment Complex lent *Top Gun* encouraged the folding of the protagonists, and by extension the viewer, into the embrace of the Petrowar machine. A constant worry for these protagonists is that they might be dismissed from the organization that allows them to exist as part of the fighter jet aircraft. Such dismissal would deprive them of access to godlike powers of destruction and for the audience, it would also mark the end of the cinematic Petrowar narrative. In addition to this, the violence that the intrepid heroes perform is clearly designed to address a crisis central to American capitalist petromodernity. In *Top Gun*, the enemy is obviously still the Soviet Union and the prize is continued access to the oil flowing from the Middle East. In this way, *Top Gun* is also an example of what Jean-Michel Valantin (2005) and Georg Löfflmann (2013) have called *national security cinema*. This is cinema that casts the 'perception of threat as an existential danger to survival, security, and order against which American power is mobilized' (Löfflmann, 2013, p. 282). As national security cinema, *Top Gun* thus poses American Petrowar as the only force capable of addressing an often vaguely described or even allegorized insecurity.

THE AMERICAN CLIMATE EMERGENCY PETROWAR NARRATIVE

While *Top Gun* is obviously a film about petro-emergency, it is not a film that clearly registers the coming of an epochal crisis for the capitalist world-ecology. Such Petrowar films do begin to appear in the late 1990s however, and as such testify to what Sarah Dunant and Roy Porter have termed the *Age of Anxiety* (1996). Dunant's and Porter's book is one of many publications circulated during the closing years of the twentieth century that note how many Americans looked towards the future with considerable trepidation. Their anxiety is fuelled, Dunant and Porter argue, by diffuse economic, technological, political, and environmental concerns that blend into a general, but similarly vague, end-of-millennium angst. This was a time of relative economic prosperity when the US had appeared to seize permanent control of the world-system, but it was also a time when the utopia promised by such development was failing to materialize. The neoliberal economic program launched by Reagan in the 1980s had begun to erode job security in the core. The IPCC, formed in 1988, had published a series of publications and their two first Assessment

Reports, published in 1990 and 1994, clearly describe the inexorable unravelling of biospheric erosion.

To the Petrowar narrative, and to the US DoD that had helped fund it, the vague contours of the anxieties that haunted the US core at this point in time were both a problem and an opportunity. Deprived of the convenient enemy the Soviet Union had constituted, the Military Entertainment Complex searched for new threats that could be vanquished on the cinema screen. As I discuss in the next chapter, the kaiju was employed as an allegorical manifestation of such a threat. However, in the mid-1990s, the most successful foils in the Petrowar film were interstellar aliens. Of these films, the most successful was *Independence Day*, directed by Roland Emmerich and opening on July 2, 1996. In the film, an armada of alien ships arrives on planet Earth on this precise date. They gather over the major metropolises of the world-system and loom threateningly over iconic constructions such as the Statue of Liberty, the Twin Towers, and the White House. On July 3, the armada launches a coordinated attack, eradicating city centres and the White House with powerful energy beams. The US military attempts to strike back, but invisible energy shields protect the alien ships, making them invulnerable to conventional Petrowar and even to atomic weapons. The American president is saved at the last minute and taken to a secret research facility where he and his advisors begin planning the resistance (Fig. 3.2).

Fig. 3.2 Caption: Energy beams from an alien ship obliterate the White House in the film *Independence Day*

Independence Day is in every way an emergency narrative. John Dixon (1996) has attributed the film's enormous commercial success to the fact that it 'appeared at a moment of widespread anxiety about job security in a culture in which downward mobility is experienced, particularly by men, as personal failure and impotence' (p. 92). In this way, it is a response to the slow unfolding of neoliberalism and the general sense of emasculation that this arguably produced for (white) men suddenly expelled from the Fordist system. In his detailed study Independence Day, *Or, How I Learned to Stop Worrying and Love the Enola Gay* (1998), Michael Rogin concurs with this general analysis but he also expands it to show how *Independence Day* responds to, and attempts to resolve, a host of anxieties that haunted US society in the late 1990s. In Rogin's reading, the film addresses fears provoked by the fact that the US was becoming increasingly multicultural and queer. These fears are then resolved by making black and Jewish people join in the attempt to save the Christian settler-state, and by reasserting the dominance of the white heterosexual. Rogin also notes how the film is part of an attempt to reaffirm the US as the good and benevolent hegemon of the world-system, a nation-state able to, and supremely interested in, defending this world-system. In this way, as the title of Rogin's study argues, the film ultimately teaches its audience to 'stop worrying' about the extractive violence that the US has exerted to reach its dominant position, and to 'love the Enola Gay'.[5]

Dixons' claim that the success of *Independence Day* is due to its effective mining of anxieties central to its particular moment in time, and to its normative assertion of US white, male, heterosexual dominance, is convincing. It is also obvious that the film is informed by the wider geopolitical developments of the era; the sense that, as Samuel Huntington (1996) puts it, the world-system is moving towards a general 'clash of civilisations' that may result, as the title of his book spells out, a 'remaking of the world order'. These anxieties are not overtly tied to socio-ecological breakdown, yet they exist, of course, as part of the inter-related epochal crisis. Again, the crisis that Moore (2015) has argued the planet is moving towards is not simply a biospheric crisis but a crisis for the capitalist world-ecology. In other words, it is a crisis experienced by

---

[5] Enola Gay is, of course, the name of the aircraft that dropped the atomic bomb over Hiroshima. To 'love the Enola Gay' is thus to love the US world-system hegemony that the bomb firmly reasserted and cemented.

and through *capitalism-in-nature* (p. 13). When this is considered, *Independence Day* allegorizes not simply a national crisis but rather the arrival of a terminal crisis for the capitalist world-ecology.

Thus, *Independence Day* can be said to give expression to what Mark Bould (2021) terms the Anthropocene unconscious. Bould's thesis is that—contrary to Amitav Ghosh's (2016) idea that virtually nobody is writing about the biospheric crisis—a great wealth of culture is informed on some level by the growing socio-ecological crisis. As Bould phrases it, 'the Anthropocene is the unconscious of "the art and literature of our time"' (p. 15). The particular political unconscious of *Independence Day* can be said to have been partly inherited from British author H. G. Wells' *The War of the Worlds* 1898), published at a moment in time when the British Empire had begun to fall apart and the US had begun to transition to the position of hegemon of the world-system. This is a story about how technologically advanced Martians have depleted their world and have come to the Earth to colonize it and establish new commodity frontiers. To make sure that readers reflect on this allegory, Wells asks them in a foreword to remember that the British exterminated the Indigenous population of Tasmania during their effort to colonize Australia. In this way, the Martian attempt to colonize the Earth enacts colonialism in reverse. The British are now exposed to the extractive violence they have for so long practised in other parts of the world. *Independence Day* leaves this foreword out. Nobody is asked to reflect on the violence of Hiroshima, Vietnam, or the 1990–1991 invasion of Iraq in *Independence Day*. Yet, this violence and this military campaign hover inverted over the violence performed by the aliens. Through its imagery of profound socio-ecological devastation, *Independence Day* still registers the extractive violence of the global, capitalist, fossil fuel economy.

This noted, it should also be observed that the solution to the apocalyptic emergency the aliens constitute in *Independence Day* is the same as in *Top Gun*: Petrowar. As in *Top Gun*, the protagonist (Will Smith) is a US Air Force pilot. Even the US president turns out to be a former pilot and in the cataclysmic final battle, he takes to the skies alongside the rest of the US war machine. Again, the only way to address the dispersed epochal crisis that the film evokes is through organized military petro-violence. In this way, *Independence Day* is in every way a fossil emergency fiction where Petrowar resolves the allegorized crisis that informs the narrative. Those who have not invested in this technology, and who do not heed the American call to a 'counter-offensive', stand impotent before the crisis.

The closure of the film is the triumphant resurrection not just of US world-system hegemony, but of *faith* in this entity.

Independence Day did seek to collaborate with the US DoD but the producers were turned down due to setting part of the movie in the mythologized Area 54 (Felber 1999). However, as argued by Löfflmann (2013), the success of the film still paved the way for the next generation of alien Petrowar films. The Hollywood spectaculars *Battle: Los Angeles* (Liebesman 2011) and *Battleship* (Berg 2012) are two examples of films that received substantial support from the US Department of Defense, and that made use of this to describe how aliens' attempts to colonize the planet are thwarted by the petro-accelerated US military. In these films, US Petrowar is again the only mechanism capable of defending a universalized humanity against the various emergencies that the alien invaders represent.[6] Thus, in fossil emergency fiction, the harrowing shape of extra-terrestrial forces obscure the politics and the planetary history that have brought about the social and ecological crises that inform the story. In this way, the US military again appears as the only entity capable of addressing the allegorized geopolitical and ecological upheaval that is at the heart of this crisis.

SETTING THE STAGE FOR THE CLIMATE EMERGENCY NARRATIVE

Many of the Fossil Fictions discussed in this chapter tell stories about how extraction depletes labour and land, but at the same time, they insist that violence is the only possible response to the crises such depletion causes. Coal miners existing on eroding commodity frontiers must be struck down by armed militia and actors within (or formally outside) the world-system that resist the expansion of a hegemonic US must be levelled with the help of Petrowar technologies. Such violence, and even more so the spectacular Fossil Fictions that narrate this violence as adventure, work to obscure the history and politics that produced it in the first place. In this way, this fiction joins the settler capitalist and plantation text in nurturing Vandana Shiva's (1993) 'monocultures of the mind' (p. 5)

---

[6] See also Johan Höglund (2014) *The American Imperial Gothic: Popular Culture, Empire, Violence.* Farnham: Ashgate.

that limit thinking capable of imagining worlds and ecological relationships beyond those produced by extractive capitalism. The coal extraction
narrative launches this type of text, and the Petrowar story further raises
the stakes of this fiction. Separated from the coal frontier narrative by
two world wars and by bloody neocolonial conflict in Asia, by US peak
oil, and by the realization that fossil fuel/capital is eroding conditions for
life on the planet, the Petrowar text is framed by a much more profound
sense of crisis and emergency. This sense is discernible in *Top Gun* and
it saturates films such as *Independence Day*. However, there is never any
doubt in these films that the US presence on the battlefields of Europe,
on the Indian Ocean, or in space, is necessary. The enormous aircraft
carrier, the constant burning of jet fuel, the rockets, machine gun fire,
and the occasional dying off of heroic pilots are all inevitable responses
to a world-system that is becoming increasingly difficult to negotiate. The
Petrowar conducted in these and similar films is the only possible response
to large-scale geopolitical (or intergalactic) emergency.

In *Top Gun*, the US is prepared for such an emergency. *Independence Day* tells a similar story, only in this film, the crisis has taken on
apocalyptic proportions. If Petrowar fails, it is not just American petromodernity that will end, but the planet. This recalls Jameson's observation
that '[i]t seems to be easier for us today to imagine the thoroughgoing
deterioration of the earth and of nature than the breakdown of late capitalism' (1994, p. xii) only what is imagined in this film is the invasion of
the planet by a hostile alien civilization. In *Independence Day*, the end of
capitalism's ability to perform Petrowar would be the end of the world.
It should also be noted that texts such as *Top Gun* are made possible
by support provided by institutions, agencies, networks, and departments
that are tasked with securing the core. DoD-sponsored national security
cinema such as *Top Gun, Battle Los Angeles,* and *Battleship* which Valantin
(2005) and Löfflmann (2013) discuss is thus core cinema in every sense of
the word. As discussed in this chapter, and as I will return to, they emerge
out of a concerted and highly conscious attempt to stimulate American
corporations, create goodwill and potential recruits for the armed forces,
and to thus extend both American hegemony of the world-system and
the capitalist world-system as such. Unlike the other texts discussed in
this book so far, they are not simply the voice of (white, male) authors
located in, and privileged by, the core: they are the booming voice of the
core.

Fossil Fiction is a type of text that builds on the foundation that the settler capitalist and plantation narratives established. This fiction notes even more clearly than the previous types that the land suffers because of extraction and other types of violence, yet it also promotes the extractive relationship to land and people as inevitable and somehow regenerative. It is through the production and telling of stories that insist that the extraction of fossil energies is vital to the US as the aspiring or dominant core of the capitalist world-system, even when such extraction causes widespread human and ecological death, that Fossil Fiction paves the way for the American Climate Emergency Narrative. Fossil Fiction is not the same thing as the American Climate Emergency Narrative since the crisis it envisioned is not epochal and ecological to the same extent as in this narrative. However, since Fossil Fiction imagines that the crises that coal and oil produce can also be addressed through the society these energy forms make possible, especially as these energy forms are militarized, it shares significant DNA with the American Climate Emergency Narrative. As coming chapters reveal, the notion that the vaguely conceived problem (extractive, capitalist modernity) is somehow also the solution, is something that the American Climate Emergency Narrative will rehearse ad nauseam.

Works Cited

Adams, Sean P. 2004. *Old Dominion, Industrial Commonwealth: Coal, Politics, and Economy in Antebellum America*. Baltimore: John Hopkins University Press.

Atkin, Nicholas., and Michael Biddiss. 2008. *Themes in Modern European History, 1890–1945*. Abingdon: Routledge.

Berg, Peter, director. 2012. *Battleship*. Universal Pictures.

Bonneuil, Christophe, and Jean-Baptiste Fressoz. 2016. *The Shock of the Anthropocene: The Earth, History and Us*. London: Verso.

Bould, M. 2021. *The Anthropocene Unconscious: Climate Catastrophe Culture*. London: Verso.

Clay, Henry. 1857. *The Life, Correspondence, and Speeches of Henry Clay*. Vol. 5. New York: A . S. Barnes & Co.

Crawford, Neta C. 2022. *The Pentagon, Climate Change, and War: Charting the Rise and Fall of US Military Emissions*. Cambridge, MA: MIT Press.

Cunningham, Rodger. 1990. 'Signs of Civilization: *The Trail of the Lonesome Pine* as Colonial Narrative.' *Journal of the Appalachian Studies Association* 2: 21–46.

Der Derian, James. 2001. *Virtuous War: Mapping the Military-Industrial-Media-Entertainment Network*. Boulder: Westview Press.
Dixon, John. 1996. 'Aliens' R'us: A Critique of D4.' *Film & History: An Interdisciplinary Journal of Film and Television Studies* 26 (1): 92–98.
Dunant, Sarah, and Roy Porter. 1996. *The Age of Anxiety*. Edited by Sarah Dunant and Roy Porter. London: Virago.
Emmerich, Roland, director. 1996. *Independence Day*. 20th Century Fox.
Feis, Herbert. 1950. *Road to Pearl Harbor*. 2015. Princeton, NJ: Princeton University Press.
Felber, Dietmer. 1999. 'Independence Day, or How I Learned to Stop Worrying and Love the Enola Gay.' *Film Criticism* 24 (1): 92–95.
Fox Jr., John. 1908. *The Trail of the Lonesome Pine*. New York: Grosset & Dunlap.
Ghosh, Amitav. 2016. *The Great Derangement: Climate Change and the Unthinkable*. Chicago: The University of Chicago Press.
Hobsbawm, Eric. 1987. *Age of Empire: 1875–1914*. London: Weidenfeld & Nicolson.
Höglund, Johan. 2008. 'Electronic Empire: Orientalism Revisited in the Military Shooter.' *Game Studies* 8 (1): 10.
———. 2012. 'Militarizing the Vampire: Underworld and the Desire of the Military Entertainment Complex.' In *Transnational and Postcolonial Vampires*, edited by Tabish Khair and Johan Höglund, 173–188. Basingstoke: Palgrave Macmillan.
———. 2014. *The American Imperial Gothic: Popular Culture, Empire, Violence*. Farnham: Ashgate.
Huntington, Samuel P. 1996. *The Clash of Civilizations and the Remaking of World Order*. New York: Simon & Schuster.
Jameson, Fredric. 1981. *The Political Unconscious: Narrative as a Socially Symbolic Act*. Ithaca, NY: Cornell University Press.
———. 1994. *The Seeds of Time*. New York: Columbia University Press.
LeMenager, Stephanie. 2014. *Living Oil: Petroleum Culture in the American Century*. Oxford: Oxford University Press.
Lenoir, Tim. 2000. 'All but War Is Simulation: The Military-Entertainment Complex.' *Configurations* 8 (3): 289–335.
Lenoir, Timothy, and Henry Lowood. 2005. 'Theaters of War: The Military-Entertainment Complex.' In *Collection, Laboratory, Theater: Scenes of Knowledge in the 17th Century*, edited by Helmar Schramm, Ludger Schwarte and Jan Lazardzig, 428–456. Berlin: De Gruyter.
Lewis, Ronald L. 1987. *Black Coal Miners in America: Race, Class, and Community Conflict, 1780–1980*. Lexington: University Press of Kentucky.
Liebesman, Jonathan, director. 2011. *Battle: Los Angeles*. Sony Pictures.

Löfflmann, Georg. 2013. 'Hollywood, the Pentagon, and the Cinematic Production of National Security.' *Critical Studies on Security* 1 (3): 280–294.

Malm, Andreas. 2016. *Fossil Capital: The Rise of Steam Power and the Roots of Global Warming*. London: Verso.

———. 2017. '"This Is the Hell That I Have Heard Of": Some Dialectical Images in Fossil Fuel Fiction'. *Forum for Modern Language Studies* 53 (2): 121–141.

Moore, Jason W. 2015. *Capitalism in the Web of Life: Ecology and the Accumulation of Capital*. New York: Verso.

Niblett, Michael. 2020. *World Literature and Ecology: The Aesthetics of Commodity Frontiers, 1890–1950*. Chamalthusser: Palgrave Macmillan.

Page, Sebastian N. 2011. 'Lincoln and Chiriquí Colonization Revisited.' *American Nineteenth Century History* 12 (3): 289–325. https://doi.org/10.1080/14664658.2011.626160.

Painter, David S. 2014. 'Oil and Geopolitics: The Oil Crises of the 1970s and the Cold War.' *Historical Social Research/Historische Sozialforschung* 39 (4): 186–208.

Robb, David L. 2004. *Operation Hollywood: How the Pentagon Shapes and Censors the Movies*. Amherst, NY: Prometheus Books.

Rogin, Michael. 1998. *Independence Day, or, How I Learned to Stop Worrying and Love the Enola Gay*. London: British Film Institute.

Sadler, Spencer J. 2009. *Pennsylvania's Coal and Iron Police*. Charleston, SC: Arcadia Publishing.

Scott, Tony, director. 1986. *Top Gun*. Paramount Pictures.

Shapiro, Henry D. 1986. *Appalachia on Our Mind: The Southern Mountains and Mountaineers in the American Consciousness, 1870–1920*. Chapel Hill, NC: University of North Carolina Press.

Shiva, Vandana. 1993. *Monocultures of the Mind: Perspectives on Biodiversity and Biotechnology*. London: Zed Books.

Shulman, Peter A. 2015. *Coal and Empire: The Birth of Energy Security in Industrial America*: Baltimore: John Hopkins University Press.

Stinnett, Robert. 2001. *Day of Deceit: The Truth About FDR and Pearl Harbor*: New York: Simon and Schuster.

Turner, Frederick Jackson. 1893. *The Significance of the Frontier in American History*. 2014. Mansfield Centre, CT: Martino Publishing.

Turse, Nick. 2008. *The Complex: How the Military Invades Our Everyday Lives*. New York: Metropolitan Books.

Valantin, Jean-Michel. 2005. *Hollywood, the Pentagon and Washington*. London: Anthem Press.

Vitale, Alex S. 2017. *The End of Policing*. London: Verso.

Wark, McKenzie. 2007. *Gamer Theory*. Cambridge: Harvard University Press.

Washington, Booker T. 1901. *Up from Slavery*. 1907. New York: Doubleday.

Wells, H. G. 1898. *The War of the Worlds*. London: William Heinemann.
Wilson, Darlene. 1995. 'The Felicitous Convergence of Mythmaking and Capital Accumulation: John Fox Jr. and the Formation of an (Other) Almost-White American Underclass.' *Journal of Appalachian Studies* 1 (1): 5–44.

CHAPTER 4

The Irradiated

CHEAP WAR AND NUCLEAR FICTIONS

As noted in Chapter 2, publications by the Anthropocene Working Group (AWG) argue that the Anthropocene begins 'historically at the moment of detonation of the Trinity A-bomb at Alamogordo in 1945' (Zalasiewicz et al. 2015, p. 200). This is because this test produced radioactive fallout that settled into the stratigraphic record where it can now be detected. An obvious problem with the AWG's dating of the formal beginning of biospheric breakdown to the Trinity test, and to the use of the atomic bomb at Hiroshima and Nagasaki, is that it ignores the material (and cultural) history that led up to the detonation of these weapons. It should be obvious that the deployment of nuclear weapons in the 1940s comes out of a specific military world-system history engineered by, and privileging, a limited group of humans. Eliding this history makes it enormously difficult to understand what has brought biospheric erosion about, and that this erosion is profoundly socio-ecological in nature rather than simply climate-related. The failure to centre this basic state of affairs also makes it difficult to prevent further erosion.[1]

---

[1] If oncology had concerned itself only with diagnosing and treating an illness such as lung cancer, and never with the reasons why this illness develops, we would still all be smoking. It should be observed that the AWG has begun to consider scholarship that connects climate breakdown to capitalism and the history of social injustice (see Gibbard et al. 2022).

© The Author(s) 2024
J. Höglund, *The American Climate Emergency Narrative*,
New Comparisons in World Literature,
https://doi.org/10.1007/978-3-031-60645-8_4

That said, while the first atomic bomb does not mark the historical beginning of the socio-ecological breakdown the AWG has called the Anthropocene, it does denote an important new material, military, economic, and cultural phase of this breakdown. Understood in relation to the long history of securitization of extraction that accompanied the evolution of the capitalist world-ecology, and thus as a particular stage of what Jason W. Moore (2016) has theorized as the Capitalocene, the deployment of the first atomic weapons began a revolution in military technology similar to what I call Petrowar in the previous chapter. While nuclear energy made it possible to further accelerate military technologies such as the submarine, its main contribution to militarized capitalism was its ability to make destruction and killing on a planetary scale *cheap*.

Robert Oppenheimer, the director of the Manhattan Project and the de-facto engineer of the atomic bomb, was deeply aware of this and commented on the bomb as precisely a *cheap* weapon after the war. In the essay 'The New Weapon: The Turn of the Screw' (1946), he estimates that in 'this past war it cost the United States about $10 a pound to deliver explosive to an enemy target. Fifty thousand tons of explosive would thus cost a billion dollars to deliver' (p. 24).[2] He then figures that the sum involved when delivering the same amount of explosive via nuclear technology is 'several hundred times less, possibly a thousand times less'. Thus, '[a]tomic explosives vastly increase the power of destruction per dollar spent, per man-hour invested' (p. 24). Oppenheimer also notes that while the nuclear bomb can be used against military personnel and equipment, 'their disproportionate power of destruction is greatest in strategic bombardment: In destroying centers of population, and population itself, and in destroying industry' (p. 25). If, as Moore (2016) and Patel and Moore (2018) have argued, capitalism has sustained itself through the cheapening of essentials such as nature, energy, food, labour, and life, the invention and use of the atomic bomb marks a new stage in this development. With the atomic bomb, planetary-scale mass death and destruction were suddenly perfectly affordable. At the same time, (settler) cities, sites of refinement, and plantations of various types had become more vulnerable than ever before.

---

[2] Robert Oppenheimer's (oddly titled) essay was part of the bestselling *One World or None* in which scientists such as Albert Einstein and Nils Bohr considered how the nuclear bomb had transformed the playing field for people, for nations, and for the planet.

This produced a new set of opportunities for the hegemon of the world-system, but also a new sense of insecurity. After the war, nuclear bomb testing was resumed under the codename Operation Crossroads. After deporting the Indigenous Micronesian population from the Bikini Atoll that is part of the Marshall Islands, the US tested a series of increasingly powerful nuclear bombs. As testing continued, and when US allies and the Soviet Union also began to test weapons, it became clear that fallout from detonations was spread across the planet, so that it could potentially enter and alter human bodies in detrimental ways. American geneticist Hermann J. Muller warned already in his 1946 Nobel Prize lecture that even minute doses of radiation may lead to genetic damage and mutations. In addition, nuclear weapons were shown to be able to drastically destroy entire geographies, and even to potentially alter the global climate. In 1952, the US erased the Pacific Island of Elugelab/Ālloklap from the face of the Earth with the help of the first hydrogen bomb 'Mike', replacing it with a radioactive crater. In 1957, the US Department of Defense published the book *The Effects of Nuclear Weapons* which speculated that nuclear detonations, just like volcanic eruptions, could be capable of throwing so much dust into the atmosphere that it may obscure light from the sun (see Glasstone and Dolan 1957, pp. 70–71). Out of this assumption emerged the notion of 'nuclear winter', described as a prolonged cooling of the planet produced by atomic warfare. In this way, the nuclear bomb was understood as a weapon that could produce slow, insidious, and mutating violence across the globe, and that may also, potentially, *change the climate of the planet*. Already in 1947, the newly formed Bulletin of the Atomic Scientists created the Doomsday Clock and set it to seven minutes to the midnight that marks oblivion.[3] With this clock firmly and publicly in place, the idea that modern society may be moving towards a terminal, military, and world-systemic crisis had been firmly established and disseminated.

This realization produced texts that speculated on the possibility that human activity might ruin the planet for human and extra-human life. Donald Worster (1985) has termed the post-war period the 'Age of Ecology' (p. ix) and argued that it was inaugurated by the 'dazzling fireball of light and a swelling mushroom cloud of radioactive gases'

---

[3] At the time of writing, in the wake of the Trump 2016–20 presidency and the Russian invasion of the Ukraine, the clock is set to 90 seconds to midnight.

(p. 342) produced by the detonation of the first atomic bomb in Alamogordo in 1945. The realization that radioactive fallout may produce mutations, that large-scale nuclear war may radically alter the conditions for life on the planet, and that the world may be moving towards a terminal ecological crisis, were registered in fiction from a US nation-state that had, at this time, clearly become the new core of the capitalist world-system/ecology. Thus, as Harold L. Berger (1976) has argued, the post-WWII period saw the birth of a new kind of science fiction that was 'prophetic, increasingly more concerned with disasters than marvels' (p. x). The event that 'fertilized the soil from which [such] science fiction grows' was, Berger observes, the 'atomic bomb that levelled Hiroshima' (p. 147). This new breed of science fiction was not overly interested in people who had, or may, suffer nuclear war in Japan though. Instead, it speculated on what life might be like for (white, middle-class) Americans in futures transformed by radiation and nuclear war. By exploring worlds where military-grade violence has damaged the planet and thrown the US nation-state into a sudden state of terminal crisis, it produced a new ecological and geological paradigm in American fiction. While the Petrowar text did evoke images of large-scale destruction, the pre-1990s Petrowar narrative tended to view this destruction as local. In the Petrowar text, there is somewhere to escape to when the land is burning. In what can be named Nuclear War Fiction, such refuge cannot be taken for granted.

This chapter discusses the irradiated worlds that these post-WWII nuclear narratives conjure. The chapter first examines the early wave of post-WWII speculative fiction that locates humans within a post-atomic war future. In some texts, this world is utopian, but in most, it has been severely damaged by nuclear military violence. Emerging out of the already existing martial and eco-phobic literary tradition described in the previous chapters, these early stories can be said to make up the earliest examples of the fully formed American Climate Emergency Narrative. As the chapter shows, the first wave of post-WWII science fiction thus registers the capacity of the militarized, modern state to profoundly alter or even destroy the precarious ecosystem of the planet. This registering, I argue, provides space for a certain radical literary tradition in science fiction, but it also paves the way for texts that cast profound ecological and social crises as, in the words of Marzec, 'an engagement opportunity' (2015, p. 9). These are texts that cast an irradiated and riotous ecology in the shape of an enormous, and enormously aggressive, chthonic being that levels megacities and that must be fought by the very military institutions that, in many cases, have participated in its creation. This type of story first appears with the release of the American *The Beast from 20,000 Fathoms* (Louri 1953) and becomes part of global popular culture with the Japanese *Gojira* (Honda 1954). Both films feature a prehistoric creature energized by nuclear energy into a planetary-scale threat. As the chapter discusses, this gigantic figure has become central to the contemporary American Climate Emergency Narrative. Particularly, as the chapter argues, the Warner Brother's Monsterverse films *Godzilla* (Edwards 2014) sponsored by the US Department of Defense, and *Godzilla II: King of the Monsters* (Dougherty 2019) are central to the evocation of the US military as the only system capable of addressing the planetary emergency it has been instrumental in creating.

Entering Irradiated Socio-Ecological Futures

When the first atomic bombs were dropped on Hiroshima and Nagasaki, the initial American reaction was triumphant. In Congress, California representative Jerry Voorhis argued that when 'the scientists released atomic energy [...] the greatest event in the history of mankind except only the birth of Christ Himself took place' (Jerry Voorhis cited by Weisgall 1994, pp. 79–80). Mississippi Representative John Rankin concurred, arguing that 'Almighty God has placed this great weapon in our hands at a time when atheistic barbarism is threatening to wipe Christianity from the face of the world' (Rankin cited by Weisgall 1994, p. 80). The detonations of the first nuclear bombs were thus presented as a kind of divine intervention, with God hailing (and securing) the progress of settler capitalism at a time of large-scale global crisis.

This jubilant greeting of the atomic bomb and the mass death it produced gave rise to a series of utopian stories such as Robert A. Heinlein's novel *Rocket Ship Galileo* (1947), and to films such as Irving Pichel's *Destination Moon* in 1950 (an adaptation of Heinlein's story), where atomic energy makes it possible to extend the settler capitalist project into space. In *Destination Moon*, a coalition of US corporations and freelancing military men build a moon-bound spaceship energized by nuclear power. The main purpose is to secure cislunar space: 'The first country that can use the moon for the launching of missiles will control the Earth! That, gentlemen, is the most important military fact of this century!' the director of the project exclaims. At the same time, the journey is conducted to ensure access to the new commodity frontiers offered within the solar system. Having reached the moon and taken 'possession' of it in the name of God and the US, the astronauts discover Uranium deposits that make the moon a viable resource for the military and also for a capitalist world-system perceived in the film as transitioning away from petromodernity towards new nuclear commodity frontiers.

However, even unapologetically utopian narratives such as *Destination Moon* voice some concern. From space, the Earth appears 'vulnerable and exposed forever', and one of the astronauts jokes that the Uranium found could be used to 'blow up the moon too'. This barely voiced counter-discourse can be traced back to the publication of John Hersey's *Hiroshima* in 1946. This horrific journalistic account of the bomb and its aftermath revealed the enormous proportions of the suffering caused by nuclear warfare. Hersey's text uses first-person testimony to piece together the massive violence released by the bomb. People, architecture, and extra-human nature closest to the detonation are vaporized by the shock wave and the firestorm, leaving an area 'of four square miles of reddish-brown scar, where nearly everything had been buffeted down and burned' (p. 21). Equally horrific is the new form of slow violence that lingers in the wake of the detonation. When the fires have been extinguished and survivors cared for, people remaining in the city of Hiroshima are struck by insidious and unexplainable illnesses. Small wounds caused by the detonation have 'suddenly opened wider and were swollen and inflamed' (p. 103), and, as if a plague of some sort has invaded the city, people vomit, their hair falls out, skin blisters, and organs fail. The land on which Hiroshima stood has been turned hostile by military violence. People sicken and die simply by entering this irradiated space.

The nightmarish images disseminated by the publication of Hersey's account merged with the aforementioned research that argues that radiation causes genetic damage and with the reports suggesting that large-scale nuclear war may alter the climate of the planet. Joseph Masco has termed what rises out of this merger 'the nuclear uncanny', described as 'a psychic effect produced, on the one hand, by living within the temporal ellipsis separating a nuclear attack and the actual end of the world, and on the other, by inhabiting an environmental space threatened by military-industrial radiation' (Masco 2006, p. 28). He furthermore argues that it is via the nuclear uncanny 'that the concept of "national security" becomes most disjointed, as citizens find themselves increasingly separated from their own senses and distrusting of their own surroundings due to an engagement with nuclear technologies' (p. 28). Put differently, the nuclear uncanny is triggered by the tacit realization that the system that is supposed to keep citizens at the core safe is in fact also eroding possibilities for life and social cohesion. Similarly, Jill E. Anderson (2021) has argued that atomic 'horror [...] actually becomes the norm that works to control the motivations of American citizens, elevating and normalizing anxieties rather than alleviating them' (p. 6). It is by exploring such problematic and divisive concerns that the first wave of post-WWII, apocalyptic texts contribute to the development of the American Climate Emergency Narrative.

As Berger contends, the Hiroshima and Nagasaki bombings, as well as continuing bomb tests in the Pacific Ocean, were followed by an outpouring of new and critical science fiction in America. Some of the first stories that describe how nuclear war produces profound and disturbing ecological havoc and thus a terminal crisis for the world-system are Theodore Sturgeon's 'Thunder and Roses' and Paul Anderson's and F. N. Waldrop's 'Tomorrow's Children', both published in 1947 in *Astounding Science Fiction*. 'Thunder and Roses' records life in a US military facility after a large-scale nuclear attack by an unnamed enemy has devastated much of the nation. The soldier protagonist notes that '[a]ll the big cities are gone. We got it from both sides. We got too much. The air is becoming radioactive' (p. 77). The question that haunts this protagonist, and the other soldiers that remain on the US base, is whether the US should retaliate or not. If they desist, the 'spark of humanity can still live and grow on this planet. It will be blown and drenched, shaken and all but extinguished, but it will live' (p. 86). If they choose to strike back, the planet will be sterilized, reduced to 'a bald thing, dead and deadly' (p. 86). The story is a striking testimony to how fiction at this time had begun to narrate military-grade nuclear violence as planetary in scale. While the Petrowar narrative discussed in the previous chapter reveals that American fiction tacitly registers militarized capitalism as destructive to ecology, the atomic bomb story further elevates military violence to an existential, planetary-scale end-game.

In Anderson's and Waldrop's 'Tomorrow's Children', large-scale nuclear war has not erased all life on the planet, but it has fundamentally altered it and also caused the collapse of most nation-states. A significant portion of the world has been transformed into spaces of 'deadness' with '[t]wisted dead trees, blowing sand, tumbled skeletons, perhaps at night a baleful blue glow of fluorescence' (p. 58). What remains of the global community is plagued by bandits, 'homeless refugees' (p. 65), radiation, and plagues. The only person who has faith in the future is General Robinson, the de-facto president of the US who is hiding out with some loyal soldiers in rural Oregon. True to his namesake, Robinson is trying to make the best of the situation, but soon discovers that radiation is transforming the human race itself: 'seventy-five per cent of all births are mutant, of which possible two-thirds are viable and presumed fertile' (p. 78). Robinson considers radical eugenic action, the obliteration of an entire mutant generation, but when his own son is born with 'rubbery tentacles terminating in boneless digits' (p. 78) instead of arms and legs, he realizes that the planet has been so thoroughly inundated by radiation that 'it's everywhere. Every breath we draw, every crumb we eat and drop we drink, every clod we walk on, the dust is there. It's in the stratosphere, clear on down to the surface, probably a good distance below' (p. 78). This is not a development that can be effectively combatted via eugenics or some other technology of capitalist modernity. The ecological crisis is terminal and humanity—and human society as Robinson imagines it—will fade away.

The Arrival of the American Climate Emergency Narrative

'Thunder and Roses' and 'Tomorrow's Children' espouse what can be described as a tentative radicalism. They are not openly critical of capitalism as a system, but they do note that the effort to securitize the world-system with the help of devastating military violence is destructive. In both short stories, the planet and the species that inhabit it have been irrevocably changed and heteronormative, capitalist society is effectively gone. In 'Thunder and Roses' continued ongoing military violence will sterilize the planet and, in 'Tomorrow's Children', the son and heir of the last US president is a tentacled hybrid. In this way, they can be considered as the first examples of narratives that consider how systemogenic violence may permanently alter the planetary biosphere. These texts are produced by white authors in close proximity to the core, but they also exemplify a tendency to critically mine the fissures and breaks that capitalism inevitably produces.

However, many of the texts that come out of this era are far less radical. In a number of fictions, the irradiated, dystopian worlds that texts such as

'Thunder and Roses' conjure are combined with the self-confident militarized storyworlds inherited from the tradition of writing begun by the settler capitalist text. In this other and very different set of texts, the mutant beings that emerge out of irradiated landscapes are not sons or daughters, but chthonic monsters. These are the stories that present the military and other members of the US security machine with what Marzec (2015) calls an 'engagement opportunity' (p. 9). Unlike the nominally radical short stories that channel the horrors of Hiroshima, these texts insist that the real emergency is not the damage done to the planet, but the damage that an eroding planet might inflict on capitalist society. In this way, this type of text can be considered the first fully formed American Climate Emergency Narrative.

The first widely circulated narrative of this type is Eugène Lourié's *The Beast from 20,000 Fathoms* (1953), loosely based on Ray Bradbury's short story 'The Fog Horn' (1951). In the film, American scientists are conducting 'Operation Experiment', the code name of a 'top-priority scientific expedition' the purpose of which is to detonate a nuclear bomb 'far north of the Arctic Circle'. On the eve of 'X-day', scientists and military personnel in the control room are excited. 'Every time one of these things goes off, I feel we're helping to write the first chapter of a new Genesis', one scientist exclaims. Yet there is also a concern: 'What the cumulative effects of these atomic explosions and tests will be, only time will tell'. Following a tense, ten-second countdown, the nuclear weapon is detonated. At this juncture, archival film from the detonation of the atomic bomb 'Baker' over the Bikini Atoll on 25 July 1946, fills the screen. In the sequence, filmed from three different perspectives, an enormous pillar of water is thrown up into the air, forming the characteristic mushroom cloud. *The Beast from 20,000 Fathoms* does not acknowledge the origin of these images, but presents them as part of its own narrative; as a test taking place in the Arctic. Even so, their inclusion makes this (and many other films that also edit them into the story) into a kind of snuff movie where images of actual human and ecological death are quietly inserted into a popular entertainment feature.[4]

---

[4] While there is no record of humans dying as a direct result of the bomb tests, the long-term effects of radiation exposure, and the death caused by such exposure, are well documented (see Wasserman and Solomon 1982). Obviously, the bombs also killed vast numbers of the flora and fauna that lived on and around the Bikini Atoll.

In the film, the effect of these inserted images on the environment is sudden and spectacular. Icebergs melt into a gaseous sea. A huge mass of earth and ice rises and crumbles into fragments. Notably, these images are eerily reminiscent of much more recent, documentary photography of the melting Arctic ice shelf. This is a sequence that clearly and presciently registers how massive, nuclear-accelerated military violence disturbs the natural ecological order (Fig. 4.1).

The dimensions of this disruption increase further when an enormous, prehistoric dinosaur eventually identified as a 'Rhedosaurus' rises out of the melted ice as an unspeakable and uncontrollable figure of planetary horror. Roger Luckhurst (2020) and Rebecca Duncan (2024) have aptly argued that, at a time of capitalogenic socio-ecological breakdown, we need to read monsters differently. As Duncan puts it, in an 'age of global but unequal crises, we might also expect a global proliferation of monsters' that speak to these crises. Duncan shows how a series of texts from the Global South 'deploys monstrosity as a language in which to dramatize experiences at the periphery, but also to render visible and interrogate the dynamic of peripheralization itself' (np). A similar kind of monstrous rendering occurs, I argue, also at the hegemonic core of the world-system. From the vantage of this core, *The Beast from 20,000 Fathoms* clearly speaks about the possibility that the militarized society of the core may produce a profound ecological and social crisis. By inserting an archival film of an atomic bomb detonation in the Pacific Ocean into a film that features an ultra-destructive, carnivorous dinosaur, the

Fig. 4.1   Collapsing Arctic icebergs in *The Beast from 20,000 Fathoms*

film creates an immediate and undeniable connection between potentially Earth System-altering military violence, and large-scale monstrosity.

With this in mind, the Rhedosaurus can be understood as an avatar of an embattled planet but also as a kind of monstrous rendering of the core itself. As such, the Rhedosaurus is an early visualization of the epochal crisis that the world-system is moving towards. The Rhedosaurus proceeds by trying to evict the militarized capitalist modernity that has awakened it from its long sleep. The dinosaur sets a course for New York and the Hudson River, its old mating grounds. Having reached Manhattan, the Rhedosaurus progresses by knocking over buildings, and by trampling and biting people and cars. The Rhedosaurus apparently hates skyscrapers and Western Union, it hates Wall Street and its pretentious architecture, and it hates the bright lights of the Coney Island amusement park. It eats police officers, military men, and other agents tasked with the securitization of the hegemonic core. Bullets do not affect the monster and when a well-aimed shot from a bazooka finally manages to penetrate its thick, prehistoric reptilian hide, the situation only worsens. Scientists discover that the 'monster is a giant germ-carrier of a horrible, virulent disease. Contact with the animal's blood can be fatal'.

As argued by Dawn Keetley (2021), this is the inauguration of another trope central to the contemporary climate narrative. The possibility that prehistoric pathogens may lie dormant in now slowly thawing permafrost is a real concern for climate science (Miner et al. 2021; Wu et al. 2022) and it informs films such as *The Thaw* (Lewis 2009) and novels such as Oana Aristide's *Under the Blue* (2021). Paving the way for such narratives, *The Beast from 20,000 Fathoms* describes how violence exerted by human activity on the Arctic permafrost releases not only enormous chthonic creatures but also a blood-borne, microscopic pathogen that has lain dormant in the ice for millennia. The virulent disease is incredibly aggressive and drops soldiers where they stand. The scientists now know what the cumulative effects of the atom bomb they set off will be: an eruption of an absolutely furious ecology that seeks to restore the violent Hobbesian order of the Cretaceous period (Fig. 4.2).

The final solution to this crisis is not, however, to cease doing violence to the planet, but instead to put renewed trust in the institution that produced this uncontrollable horror in the first place. The 'full-scale war' that the National Guard fights on the streets of New York may not have the desired effect, but when the military again turns to the nuclear option, a viable solution is discovered. A 'radioactive isotope' is shot into the

Fig. 4.2   The Rhedosaurus on Wall Street in *The Beast from 20,000 Fathoms*

monster's bloodstream, killing it, and 'destroying all that diseased tissue'. The beast thrashes and screams in front of a burning Coney Island roller coaster until the film ends. Militarized atomic energy is restored to its sublime state and the monster and its payload of pathogens are sent back into oblivion by the same energy, the same military complex, and the same logic as that which raised it from the dead.[5]

The Beast from 20,000 Fathoms creates a template for the (cinematic) Climate Emergency Narrative where gigantic monsters rise in response to an ecological disruption caused by military violence. In the 1950s, it was followed most immediately by *Them* (1954), a cult film where ants

[5] To Mark Janovich (1996) this means that the film furtively promotes a scientific, rather than a conventional military, solution (52). However, in this as in most other films of this type, science in general, and nuclear science in particular, is so entangled with the military industrial complex that it does not truly exist outside of it. The radioactive isotope is not so much a work of science as a nuclear weapon created for the armed forces.

in the Alamogordo desert (where the first atomic bomb was tested) have grown enormous and begun to attack the nearby community. In *It Came from Beneath the Sea* (1955) a gigantic octopus has been driven from its habitat by nuclear bomb testing, and in the British *The Giant Behemoth* (1959) a monster very similar to the Rhedosaurus from *The Beast from 20,000 Fathoms* destroys much of London.

However, the most well-known of the many monster films from this era is the Japanese *Gojira* (1954).[6] This version of the gigantic and irradiated monster—typically referred to as *kaiju*—drew from *The Beast from 20,000 Fathoms*, but it was also inspired by US testing of the thermonuclear bomb Castle Bravo on 1 March 1954. This test yielded 15 megatons, 2.5 times the original estimate and almost 1000 times that of the Hiroshima bomb (Brown 2014). The energy released produced extensive fallout as radioactive and pulverized coral was thrown up into the atmosphere and spread throughout the surrounding sea. In the days that followed radioactive particles were detected in many parts of the world (Bouville 2020). André Bouville (2020) estimates that nuclear bomb tests such as Castle Bravo caused tens of thousands of cases of cancer in the US alone, but the fallout from Castle Bravo first created acute radiation sickness in the population inhabiting many Pacific Islands and also in the crew of a Japanese fishing vessel called the *Daigo Fukuryū Maru*, or *Lucky Dragon* (Cho 2019). When the ship returned to Japan with a severely ill crew and a cargo of thoroughly radiated and inedible tuna, the nation was again forcibly reminded of the bombings of Hiroshima and Nagasaki. Before the year was over, Ishirō Honda had written, directed, and premiered the first of many films featuring this particular reptilian avatar of the planet.

Just like *The Beast from 20,000 Fathoms, Gojira* is a consequence of nuclear bomb testing. Scientists in the film speculate that radiation from nuclear bombs has mutated an ancient sea creature into a 50-foot-tall monster that does not only sink ships and topple buildings with the help of its incredible bulk, but also breathes atomic fire out of its mouth. When Japanese forces build a large electric fence to prevent it from reaching Tokyo, Gojira melts it with its atomic breath. The monster then wanders into the city, wading through architecture and terrified civilians. Just like the Rhedosaurus, Gojira appears to hate the modernity that has, accidentally, spawned it. The death toll is enormous and when Gojira escapes

---

[6] In what follows, I refer to the Japanese film, and to the monster of the Japanese franchise, as 'Gojira', and to the US version of the same creature as 'Godzilla'.

back into Tokyo Bay, survivors suffer, as did those who lived through Hiroshima and Nagasaki, from horrible radiation poisoning. Mark Bould has suggested in *The Anthropocene Unconscious* (2021) that 'Godzilla is the [atomic] bomb' (p. 27), but a more precise analysis is that Gojira, like the Rhedosaurus, is the *planet*. These two monsters thus manifest how the planet reacts to the enormous military-grade violence that the atomic bomb unleashes.

Gojira is a fiction from the Asian semiperiphery of the world-system, and as such notably different from *The Beast from 20,000 Fathoms*. While the scientists of the American film very quickly understand what the Rhedosaurus is, how it came to be, where it is heading, and why, Gojira's origins and purposes remain mysterious in the Japanese film. In addition to this, Gojira is much larger and far more devastating than the American counterpart. Like the monster texts analysed by Duncan, *Gojira* can be said to register the experience of being (semi)peripheralized. This is an experience that takes many shapes (economic, military, medical, social), but it obviously needs to be traced back to the moment when the first bombs were dropped on Japanese soil, when Japan lost its bid for world-system hegemony, and entire cities were wiped from the land. It must also be connected to the lingering slow violence inherent in the encounter with irradiated clouds and contaminated fish in the wake of nuclear bomb tests carelessly performed by the victorious and dominant core. It is indeed not strange that Gojira renders the experience of being semi-peripherialized as that of being trampled, having cities levelled, and being washed by atomic fire. In this way, *Gojira* can be described as a kind of translation of *The Beast from 20,000 Fathoms* into Japanese for a traumatized and peripheralized audience.

For the rest of the twentieth century, there is a constant, low-budget production of kaiju movies in both the US and Japan. These two branches have some things in common. The irradiated monster rises out of seas, ice vistas, or the perforated underbelly of the planet to invade the vast urban cityscapes capitalist modernity has created. Iconic buildings are scaled and crushed, and soldiers, guns, and tanks are trampled. However, in *Gojira* and many of the Japanese sequels, the violence is far more extensive, unexpected, sudden, and spectacular than in the American, drive-in examples. In the American monster narrative, the emergence of an irradiated, gigantic, but also righteously angry, avenger makes a different kind of sense. There is a certain radical logic to the arrival of the Rhedosaurus on Wall Street. The notion of a gigantic, prehistoric being awakened by the testing of nuclear weapons and returning to the city and the context which brought it to life testifies to the surfacing of a political (or 'Anthropocene', as Bould would have it) unconscious. These irradiated gigantic creatures tacitly express the fear that the extractive violence performed in the peripheries of the world-system—the Alamogordo desert, the Arctic, or the Indigenous Pacific—may come back to literally bite the capitalist core that has organized it. Read in this way, the Rhedosaurus is a figure that represents an emerging climate emergency. It is a creature let loose by melting Arctic ice, but it can also be understood as the crisis itself. When the Rhedosaurus invades Wall Street it is bringing the crisis it represents to the location most symbolic of the core within the world-system.

However, as Treasa De Loughry (2020) has observed, fiction from the core is strikingly adept at containing such eruptions of the political unconscious. Thus, while the American post-WWII monster spectacular does register the advent of a new type of economic and ecological crisis for capitalism, it also finds ways to resolve this entangled crisis. To this effect, most US versions pose an alliance between nuclear science and the military-industrial Petrowar complex as the only viable solution to the imminent emergency the monster constitutes. The ending of the US version, up until the turn of the millennium, is invariably the death of the monster at the hands of the military and the military science complex. The triumphant reestablishment of capitalist modernity follows. In this way, the solution to the emergency these texts evoke is more of the violence that produced the emergency in the first place. Ultimately, the object of this first version of the American Climate Emergency Narrative is not to restore *ecology* but the *conditions that allow capitalism to perform its extractive work*. Thus, the closure of the American kaiju film is the restoration of the state of business-as-usual. To the uniformly white, middle-class people who run to shelter when the kaiju appears, and who are ultimately saved by the intervention of technoscience, this ending is utopian and greeted by triumphant music. The suffix 'The End' that completes the film ultimately signals the return of this particular state of capitalist equilibrium.

A New Godzilla Rises

Except for the 1976 remake of *King Kong* directed by John Guillermin, the monster spectacular entered the 1970s as a low-budget, fringe culture phenomenon. It was not until 1998 that advances in CGI technology, along with French nuclear bomb testing in Polynesia, prompted Tristar Pictures to create the first mega-budget US version of the Godzilla story. As in *The Beast from 20,000 Fathoms*, the original Japanese *Gojira* and their various copies, what brings this Godzilla into the world is the release of militarized nuclear energy, although in this case, these tests are performed by the French. Even so, and in concert with previous films, military violence is perceived as the ultimate antidote to the ecological/military crisis that Godzilla represents. Trapped by the steel suspension cables of Brooklyn Bridge, Godzilla is an easy target for the fighter jets sent to end the monster's sudden incursion.

While the 1998 *Godzilla* makes use of new-generation CGI to create a uniquely realistic-looking monster, it retains much of the (partially involuntary) comedy that saturates the early kaiju film. The film was shot in the wake of what Francis Fukuyama termed the *End of History* (1992), a time when liberal capitalist society appeared to have eclipsed all other forms of social order, when the threat of Soviet nuclear holocaust seemed all but over and when China was little more than a new periphery to source for cheap labour. But, as discussed in the previous chapter, the film also comes out of what has been called the *Age of Anxiety* (Dunant and Porter 1996), a time when the IPCC had produced a first series of reports showing how the climate was quickly warming, and when it was also becoming increasingly clear that capitalist predominance did not mean that democratic principles and core privileges would become global. Giovanni Arrighi was one of a growing chorus of voices that predicted that, in the years to come, extractive, capitalist 'violence in the world system at large will get even more out of control than it already has, thereby creating unmanageable law and order problems for capital accumulation on a world scale' (Arrighi 1994, p. 342). Registering these different historical, social, and economic developments, the 1998 *Godzilla* does not seem to know if the crisis it depicts is a joke or not.

When *Godzilla* was rebooted in 2014 by Legendary Pictures and Warner Bros, it entered, and registered, a world where many of the anxieties voiced before the turn of the millennium had been realized, and where a host of new insecurities had emerged. The terrorist attacks of 9/11, the both slow and fast violence that followed the invasions and occupations of Afghanistan in 2001 and of Iraq in 2003, and the global recession that began in 2007 and that was a consequence of these wars and of the unsustainability of the capitalist project, left many in America and across the world-system reeling.[7] The resurrection of Russia as a capitalist competitor seemingly rich in cheap nature and energy, and the rise of China and India as fast-growing economic superpowers, further added to the increasing sense of vertiginous crisis. If the pre-millennial period had promised the end of all histories except that of liberal capitalism, the post-millennial era appeared to give credence both to Arrighi's warning of increasing world-system tension, and to Samuel P. Huntington's (1996) aforementioned suggestion that the coming years would see a prolonged 'clash of civilisations'. Radical intellectuals (Chomsky 2003; Hardt and Negri 2000; Harvey 2003; Johnson 2000; Wallerstein 2003) who had long recognized that the US was effectively a capitalist empire, suddenly found their thesis supported by neoconservative, right-wing pundits and historians who called American imperialism the world's best and last hope for stability, and urged continuing US military expansion (Boot 2001; Ferguson 2005). The notion of an American Empire thus arrived in the general political conversation at a time when this entity appeared to be heading for its inevitable fall.

This sense of political, military, and economic crisis combined with new climate research that conclusively showed that the planet was facing an ecological emergency of existential dimensions. This understanding had been growing at the core since the 'Age of Ecology', but it was importantly energized and transformed by the publication of Crutzen and Stoermer's aforementioned article 'The "Anthropocene"' in the May 2000 newsletter of The International Geosphere–Biosphere Programme (IGBP). As discussed in the introduction, this concept has important shortcomings, but, as Moore notes, it was also a concept that importantly sounded 'the alarm' (Moore 2016, p. 6), and made an increasingly large number of people (in the Global North) aware of the fact that human activity is the most important factor for the development of the climate of the Earth System. During the first few years of the millennium, the idea that the climate was transforming was evidenced by a series of natural disasters 'supercharged' (Rice et al. 2022) by climate change. These had the most devastating effects on precarious communities in the Global South, but events such as Hurricane Katrina in 2005 and the 2011 Texas wildfires took place in close proximity to the core and clearly impacted lives lived there. In this way, the erosion of the Earth System,

---

[7] Moore terms this recession a 'signal crisis' (2015, p. 13) for capitalism, that is, a crisis that signals the onset of terminal capitalist crisis.

the depletion of soil and of Cheap Nature generally, and the dire conditions that such erosion and depletion caused for the extractive capitalist world-system began to be perceived precisely as a profound, dispersed, and interrelated *emergency.*

As this book argues, this emergency and the layered insecurities that it is made of were essential to the American Climate Emergency Narrative after the turn of the millennium. When the next generation of American kaiju cinema appeared, it did so deeply informed by these precise crises. In particular, *Godzilla* and its 2019 sequel *Godzilla: King of the Monsters* gave form to this post-millennial trend. More than in previous US monster spectaculars, the monsters of these two films evoke a sense of terminal, epochal, and planetary crisis. Whereas the Rhedosaurus and the original Japanese Gojira confined their movement to a single city, the new Godzilla, and the other kaiju that feature in the films, move very quickly across (and through) much of the planet and the audience is repeatedly told that total human extinction is a definite possibility. Furthermore, both new films explicitly tie the enormous monsters they showcase to climate breakdown while being at the same time clearly aware of the history of irradiation and military violence that has produced the ongoing crisis.

Thus, in *Godzilla: King of the Monsters*, the eponymous monster, and the other mountain-sized kaiju (referred to as 'Titans' in the film) that feature in the re-awakened franchise, are clearly described as having been provoked into being by what the film describes as anthropogenic climate change. One of the many scientists who feature in the film explains in a sad voice to both fellow scientists within the film and to the cinema audience that:

> Humans have been the dominant species for thousands of years, and look what's happened. Overpopulation. Pollution. War. The mass extinction we feared has already begun. And we are the cause. We are the infection. But like all living organisms, the Earth unleashed a fever to fight this infection. Its original and rightful rulers. The Titans. They are part of the Earth's natural defence system. A way to protect the planet, to maintain its balance. But if governments are allowed to contain them, destroy them, or use them for war, the human infection will only continue to spread. And within our lifetime, our planet will perish. And so will we. Unless we restore balance.

In this way, Godzilla and other kaiju are described as extensions of an ecosystem that has been severely damaged by a universalized humanity that has overpopulated, polluted, and made war on the planet. Clearly, as is the scientist's point, they cannot be fought in the conventional sense, with the help of guns, rockets, nukes, fighter jets, submarines, and special forces soldiers. In fact, fighting them is the opposite of what humans should be doing.

This seems to steer the narrative away from the logic of environmentality: the militarized mentality that, as Robert P. Marzec (2015) observes, proposes that an eroding environment presents the US DoD with an 'engagement opportunity' (p. 9).[8] Logically, with the scientist's analysis in mind, the audience should expect Godzilla and the other Titans to level cities, to cull the human horde, to de-industrialize, and to disarm global society. How else can the Titans restore a balance clearly destroyed by human (militarized and capitalist) society? But *Godzilla* is fiction from the core in the same way as *Top Gun* (1986) and *Battle: Los Angeles* (2011) discussed in the previous chapter.[9] Supported by the US Department of Defense in exchange for a narrative that furthers the ideals and priorities of the DoD, *Godzilla* cannot tell a story where US soldiers, fighter jet pilots, and billions in military hardware do not help to secure the nation. Energized by the DoD, these films must find a way for modernity to dismantle the problematic eco-logic that informs the scientist's narrative.

In *Godzilla*, this is done by creating a new category of monster that *does* try to evict humanity. In *Godzilla*, this creature is referred to as the *Massive Unidentified Terrestrial Organism or* MUTO. This monster finds

---

[8] In fact, Gareth Edwards, the director of the 2014 *Godzilla*, told a story about the futility of combating ecology in his first film *Monsters* (2010). In this award-winning yet extremely low-budget movie, starring several amateur actors from the Global South, enormous alien beings have arrived via asteroids and merged with the ecosystem in parts of tropical South America. To keep these beings out of the US, the US DoD has erected enormous walls and keeps bombing the jungle within which the creatures live. However, this border thinking and militarized violence accomplishes very little in the film. The world is porous and tentacular, and the bomb runs and walls come across as nihilistic exercises in impotent power rather than as a constructive way of restoring an unsustainable social and ecological past.

[9] To be precise, *Godzilla* (2014) applied for and did receive support from the DoD (see Secker). *Godzilla: King of the Monsters* (2019) was apparently turned down by the DoD and as such it tells a slightly different story about Godzilla's origins. That noted, the 2019 sequel still mostly follows in the military tracks laid down by *Godzilla*.

and consumes radioactive material in nuclear power plants and in stock-piles of nuclear weapons, this as a preliminary to laying thousands of eggs that will produce an army of MUTOs so vast that organized capitalist modernity must surely go extinct. Before long, Godzilla is fighting these breeding kaiju instead of the humans that are arguably destroying the planet. Although not receiving support from the DoD, *Godzilla: King of the Monsters* follows suit. In this film, the angry avatar of the planet is King Ghidorah, an enormous, flying, three-headed kaiju. However, it turns out that he is not an eco-restorative Titan, but an invasive species from outer space. He challenges Godzilla as the new alpha of the Titans and initiates a massive, terraforming project that will make the Earth unin-habitable for humans, transforming the planet in ways strikingly similar to those of extractive capitalism. Again, Godzilla and allied Titans proceed to fight not the humans that are described as having disrupted the Earth System, but instead the kaiju that attempts the terraforming project. This Godzilla, it turns out, does not hate humans, houses, cities, modernity, power lines, trains, Wall Street, and the military.What this Godzilla hates are the MUTOs and King Ghidorah as they lay waste to human cities or when they bring volcanos to such frenzied eruptions that the biosphere begins to transform. Meanwhile, the scientist who has warned her fellows and the US military machine that the Titans are avatars of the planet, and that they must be released (rather than experimented on), has been outed as part of an eco-terrorist endeavour. While her analysis stands, her misguided effort to aid the Titans has caused additional harm to the planet (Fig. 4.3).

The military cleverly adapts to this shifting scenery by joining forces with Godzilla who is now an (expendable) extension of the military itself. In *Godzilla*, US warships and planes escort this new asset as it heads towards its final showdown with the MUTOs. The protagonist of the film is white, heroic US Navy officer Ford Brody who has just returned from Middle Eastern battlegrounds where he has been dismantling improvised explosive devices. Along with his fellow soldiers, he promises to do 'any-thing it takes' to stop the hostile species. For no good reason except to flaunt military technology, his platoon conducts a high-altitude (HALO), night-time parachute drop into San Francisco where the battle is already raging. While Godzilla keeps the MUTOs busy, Brody conducts conven-tional Petrowar on the hatching brood, burning them with gasoline from a fuel truck that has crashed into the now-exposed underbelly of the city. The brood expires in flames and thanks to the distraction the burning

Fig. 4.3 King Ghidorah terraforming the Earth and fighting the US military in *Godzilla: King of the Monsters*

eggs cause, Godzilla is able to tear the MUTOs to shreds, breathing arcane atomic fire into their bodies. Brody unites with his burgeoning family and Godzilla returns to the sea. *Godzilla: King of the Monsters* tells an identical story, where the entire arsenal of the militarized state—nuclear submarines, fighter jets, helicopters—are again showcased and used to wage war on King Ghidora. When Godzilla falters, the military energizes the monster by detonating a nuclear weapon in the kaiju's immediate vicinity. Godzilla hunts Ghidorah down and again breathes devastating atomic energy into the invasive monster's body.

In this way, both films can be seen to perform a (disrespectful) reworking of the original Gojira figure into an avatar, not of the planet, but of humanity or modernity or even the military. This poorly executed sleight of hand brings socio-ecological breakdown to the surface but only so that it can be appended to a story about the vitality of militarized capitalism. Again, when the entire US military machine is depicted as releasing a storm of missiles, depleted uranium bullets, and shells from tanks and warships into immense monsters in a Hollywood film supported by the DoD, what emerges out of the ruins, smoking monster bodies and raging fires is not ecology victorious but the notion that ecology can be successfully combatted. Militarized nuclear energies may produce monsters, but it is also somehow the antidote to ecological monstrosity.

ATOMIC EMERGENCIES

The arrival of the atomic age further evolved the story told by the settler capitalist story, the plantation narrative, the coal frontier text, and Petrowar fiction. The enormous destructive potential of the nuclear bomb suggested the very real possibility that the militarization of the world-system could lead to planetary-scale ecological demise. The launching of the Doomsday Clock by the Bulletin of the Atomic Scientists even located a moment in the near future when human military conflict might have altered the Earth System to such an extent that life, as it was lived in the core by certain strata, would be impossible. This prompted the publication of several novels and short stories that register this possibility and contemplate solutions to it. Some of this writing is nominally radical and ignores the Cold War politics that stated that mutual destruction was preferable to yielding control of the world-system. The same development also prompted monster spectaculars where an abused and irradiated ecology takes the form of a gigantic lizard or insect. Thus, films such as *The Beast from 20,000 Fathoms* tell stories of how the release of militarized energies literally produces monsters who then proceed to wreak vengeance on the militarized modernity that created it.

In this way, stories taking place in irradiated worlds register the ecological violence that militarized capitalism exerts. When people living through a future atomic war decide to die in bunkers rather than retaliate and thus sterilize the entire planet, or when they bring tentacular offspring into the world, it becomes possible to understand and critique the ecocidal practices of the US military on a very basic level. When an abused planet rises in the form of gigantic insects or an irradiated lizard to destroy the capitols of the world-system and to lay waste to the war machines ostensibly built to protect them, this allegory enables a certain critical understanding of the history of violence and the policies that have produced escalating socio-ecological breakdown. As *The Beast from 20,000 Fathoms* suggests, the irradiated monster is a logical consequence, as well as a just response, to violence performed on ecology by the military-industrial complex.

It must also be noted that the emergency that the kaiju narrative registers is not simply ecological, but profoundly systemic. Again, and as Moore (2015) has proposed, the current crisis is combined in the sense that the capitalist world-ecology's depletion of the planet's capacity to yield nature cheaply is also manifesting as a crisis for the capitalist world-system as such. This system has for some time undergone 'an irreversible decline in capital's capacity to restructure its way out of great crises' (p. 35). This means, as discussed in the introduction, that the world-system 'may be experiencing not merely a transition from one phase of capitalism to another, but something more epochal: the breakdown of the strategies and relations that have sustained capital accumulation over the past five centuries' (p. 13). In other words, the planetary ecological disaster that extractive capitalism has caused is becoming a crisis not simply for those humans and extra-humans inhabiting the commodity peripheries and semiperipheries where resources are extracted (and atomic bombs tested), or even for the US as the core of the capitalist world-ecology, but for the capitalist world-ecology as such.

The planetary proportions of the kaiju, their obvious hatred of the various megacities of the capitalist world-system, and the fact that they so clearly constitute an extinction-level threat to human capitalist society make them emblematic of this development. They first appear at the dawn of the Age of Ecology, at the very moment that the Great Acceleration begins saturating the planet with carbon and radioactive fallout, and in this early form, they denote the possibility that the now dominant, extractive, and militarized capitalist world-system directed by the US may just be too extractive and effective for the planet to endure. When the kaiju reappears just before the millennium, and when it becomes a central figure on the cinematic stage in the 2010s and 2020s, it does so as an avatar of the planet but also as a figure representing a social and economic crisis so profound that the world-ecology may not survive it. Since the first kaiju appears in *The Beast from 20,000 Fathoms,* this monster has grown so that the sheer size of the kaiju signals the similarly enormous scale of the crisis that these texts register.

That said, *The Beast from 20,000 Fathoms*, *Godzilla*, and *Godzilla: King of the Monsters*, are fictions produced at the very core of the world-system. They may all declare the arrival of an imminent and colossal emergency, but with the same straight face as they declare the crisis, they also propose that the only imaginable solution to this epochal crisis is more of the institutions and processes that produced it in the first place. In this way, the US kaiju film seeks to contain the crisis it narrates and to structure how audiences understand the stakes of this crisis. Hijacked by the DoD, the potential of the kaiju story to problematize the long history of military and capitalist ecological violence is almost entirely fore-closed. At the same time, the terminal ecological and systemic crisis of the present becomes little more than an occasion to demonstrate the deter-mination and arsenal of the military. Indeed, as I have argued, American (DoD-sponsored) kaiju films powerfully illustrate Marzec's observation in *Militarizing the Environment* that 'nature comes into existence in the narration of the security state as fundamentally a concern of the war machine, and catastrophic socio-ecological breakdown is taken as an "engagement opportunity"' (p. 9). In the kaiju story from the core, the military is given an opportunity to engage ecology gone wild and to beat it back into the planet.

Works Cited

Anderson, Jill. 2021. *Homemaking for the Apocalypse: Domesticating Horror in Atomic Age Literature & Media*. New York: Routledge.

Anderson, Paul, and F.N. Waldrop. 1947. 'Tomorrow's Children.' *Astounding Science Fiction* 39: 56–79.

Aristide, Oana. 2021. *Under the Blue*. London: Serpent's Tail.

Arrighi, Giovanni. 1994. *The Long Twentieth Century: Money, Power, and the Origins of Our Times*. London: Verso.

Berger, Harold. 1976. *Science Fiction and the New Dark Age*. Bowling Green: Bowling Green University Popular Press.

Boot, Max. 2001. 'The Case for American Empire: The Most Realistic Response to Terrorism Is for America to Embrace Its Imperial Role.' *The Weekly Standard*, October 14.

Bould, Mark. 2021. *The Anthropocene Unconscious: Climate Catastrophe Culture*. London: Verso.

Bouville, André. 2020. 'Fallout from Nuclear Weapons Tests: Environmental, Health, Political, and Sociological Considerations.' *Health Physics* 118 (4): 360–381.

Bradbury, Ray. 1951. 'The Fog Horn.' *The Saturday Evening Post*, June 23.

Brown, April L. 2014. 'No Promised Land: The Shared Legacy of the Castle Bravo Nuclear Test.' *Arms Control Today* 44 (2): 40–44.

Cho, Yu-Fang Cho. 2019. 'Remembering Lucky Dragon, Re-membering Bikini: Worlding the Anthropocene Through Transpacific Nuclear Modernity.' *Cultural Studies* 33 (1): 122–146.

Chomsky, Noam. 2003. *Hegemony or Survival: America's Quest for Global Dominance*. New York: Metropolitan Books.

Crutzen, Paul and Eugene Stoermer. 2000. 'The "Anthropocene".' *IGBP Newsletter* 41: 17–18. http://www.igbp.net/download/18.316f18321323 470177580001401/1376383088452/NL41.pdf.

De Loughry, Treasa. 2020. *The Global Novel and Capitalism in Crisis*. Chamalthusser: Palgrave Macmillan

Dougherty, Michael, director. 2019. *Godzilla: King of the Monsters*. Warner Bros. Pictures.

Dunant, Sarah, and Roy Porter. 1996. *The Age of Anxiety*, edited by Sarah Dunant and Roy Porter. London: Virago.

Duncan, Rebecca. 2024. 'Geography: Multiscalar Monsters.' In *A Cultural History of Monsters: Modern Monsters*, edited by Jeffrey Andrew Weinstock. London: Bloomsbury Academic.

Edwards, Gareth, director. 2010. *Monsters*. Vertigo Films.

———. 2014. *Godzilla*. Warner Bros. Pictures.

Ferguson, Niall. 2005. *Colossus: The Rise and Fall of the American Empire*. London: Penguin.

Fukuyama, Francis. 1992. *The End of History and the Last Man*. New York: Free Press.

Gibbard, Philip L., Andrew M. Bauer, Matthew Edgeworth, William F. Ruddiman, Jacquelyn L. Gill, Dorothy J. Merritts, Stanley C. Finney, Lucy E. Edwards, Michael JC. Walker, and Mark Maslin. 2022. 'A Practical Solution: The Anthropocene Is a Geological Event, Not a Formal Epoch.' *Episodes Journal of International Geoscience* 45 (4): 349–357.

Glasstone, Samuel, and Philip J. Dolan. 1957. *The Effects of Nuclear Weapons*. Washington: US Department of Defense.

Hardt, Michael, and Antonio Negri. 2000. *Empire*. Cambridge: Harvard University Press.

Harvey, David. 2003. *The New Imperialism*. Oxford: Oxford University Press.

Heinlein, Robert A. 1947. *Rocket Ship Galileo*. New York: Scribner's.

Hersey, John. 1946. *Hiroshima*. New York: Alfred A Knopf.

Honda, Ishirō, director. 1954. *Gojira*. Toho.

Huntington, Samuel P. 1996. *The Clash of Civilizations and the Remaking of World Order*. New York: Simon & Schuster.

Jancovich, Mark. 1996. *Rational Fears: American Horror in the 1950s*. Manchester: Manchester University Press.

Johnson, Chalmers. 2000. *Blowback: The Costs and Consequences of American Empire*. New York: Metropolitan Books.

Keetley, Dawn. 2021. 'Climate Change, "Anthropocene Unburials" and Agency on a Thawing Planet.' *Science Fiction Film and Television* 14 (3): 375–393.

Lewis, Mark A., director. 2009. *The Thaw*. Lionsgate.

Louri, Eugène, director. 1953. *The Beast from 20,000 Fathoms*. Warner Bros.

Luckhurst, Roger. 2020. 'After Monster Theory? Gareth Edwards's Monsters.' *Science Fiction Film and Television* 13 (2): 269–290.

Marzec, Robert P. 2015. *Militarizing the Environment: Climate Change and the Security State*. Minneapolis University of Minnesota Press.

Masco, Joseph. 2006. *The Nuclear Borderlands*. Princeton: Princeton University Press.

Miner, Kimberley R., Juliana D'Andrilli, Rachel Mackelprang, Arwyn Edwards, Michael J. Malaska, Mark P. Waldrop, and Charles E. Miller. 2021. 'Emergent Biogeochemical Risks from Arctic Permafrost Degradation.' *Nature Climate Change* 11 (10): 809–819.

Moore, Jason W. 2015. *Capitalism in the Web of Life: Ecology and the Accumulation of Capital*. New York: Verso.

———. 2016. 'Introduction: Anthropocene or Capitalocene? Nature, History, and the Crisis of Capitalism.' In *Anthropocene or Capitalocene: Nature, History, and the Crisis of Capitalism*, edited by Jason W. Moore, 1–11. Oakland, CA: PM Press.

Muller, Hermann J. 1946. 'The Production of Mutations.' December 12. https://www.nobelprize.org/prizes/medicine/1946/muller/lecture/.

Oppenheimer, Robert. 1946. 'The New Weapon: The Turn of the Screw.' In *One World or None*, edited by Dexter Masters and Katharine Way, 22–25. New York: McGraw Hill.

Patel, Raj, and Jason W. Moore. 2018. *A History of the World in Seven Cheap Things: A Guide to Capitalism, Nature, and the Future of the Planet*. London: Verso.

Pichel, Irving, director. 1950. *Destination Moon*. Eagle-Lion Classics.

Rice, Martin, Lesley Hughes, Will Steffen, Simon Bradshaw, Hilary Bambrick, Nicki Hutley, Dinah Arndt, Annika Dean, and Wesley Morgan. 2022. *A Supercharged Climate: Rain Bombs, Flash Flooding and Destruction*. Climate Council of Australia. https://www.climatecouncil.org.au/wp-content/upl oads/2022/03/Final_Embargoed-Copy_Flooding-A-Supercharged-Climate_ Climate-Council_ILedit_220310.pdf.

Secker, Tom. (n.d.). 'DoD Production Agreement for Godzilla (2014)' Blog Post at *Spy Culture*. https://www.spyculture.com/dod-production-agreement-for-godzilla-2014/.

Sturgeon, Theodore. 1947. 'Thunder and Roses.' *Astounding Science Fiction* XL (3): 76–96.

Wallerstein, Immanuel. 2003. *The Decline of American Power: The US in a Chaotic World*. New York: New Press.

Wasserman, Harvey, and Norman Solomon. 1982. *Killing our Own, the Disaster of America's Experience with Atomic Radiation*. New York: Delta Books.

Weisgall, Jonathan M. 1994. *Operation Crossroads: The Atomic Tests at Bikini Atoll*. Annapolis: Naval Inst Press.

Worster, Donald. 1977 [1985]. *Nature's Economy: A History of Ecological Ideas*. Cambridge: Cambridge University Press.

Wu, Ruonan, Gareth Trubl, Neslihan Taş, and Janet K. Jansson. 2022. 'Permafrost as a Potential Pathogen Reservoir.' *One Earth* 5 (4): 351–360.

Zalasiewicz, Jan, Colin N. Waters, Mark Williams, Anthony D. Barnosky, Alejandro Cearreta, Paul Crutzen, Erle Ellis, Michael A. Ellis, Ian J. Fairchild, and Jacques Grinevald. 2015. 'When Did the Anthropocene Begin? A Mid-Twentieth Century Boundary Level Is Stratigraphically Optimal.' *Quaternary International* 383: 196–203.

CHAPTER 5

Geopolitics

GEOPOLITICS IN THE CAPITALOCENE

The history explored so far in this book reveals that war has been a way for actors within the capitalist world-system to manage the needs and the emergencies that are endemic to capitalism. War has thus been used to ensure access to commodity frontiers located on Indigenous lands, to control labour, and to gain or maintain hegemonic positions within the world-system. A point here is that this essentially violent relationality has contributed greatly to the erosion of profoundly nestled human social and ecological worlds. As discussed in Chapter 3, it is this development that prompts Christophe Bonneuil and Jean-Baptiste Fressoz (2016) to name the current climate period the *Thanatocene*.[1] Considering the immanent need for collaboration across state, national, and continental borders to address the interrelated biospheric and socioeconomic crisis, it is vital to put an end to the 'natural history of destruction' (Bonneuil and Fressoz 2016, p. 124) that has characterized the Thanatocene, yet war remains a crucial geopolitical mechanism within the capitalist world-ecology. Although employing a Foucauldian (rather than a materialist) register, this is Robert P. Marzec's important point in *Militarizing the*

---

[1] Justin McBrien (2016) turns from Greek to Latin to name the current climate era the *Necrocene*, but his focus is also the violent extinction of life that the pursuit of Cheap Nature produces.

© The Author(s) 2024
J. Höglund, *The American Climate Emergency Narrative*,
New Comparisons in World Literature,
https://doi.org/10.1007/978-3-031-60645-8_5

Environment (2015). As he describes, the current socio-ecological crisis is frequently approached and acted upon via 'a militarized mentality, one that commandeers a consciousness to wholly rethink and replace a rich, complex, multinarrative environmental history with a single ecosecurity imaginary for the post–Cold War, post-9/11 occasion' (p. 4). Steered by this mentality, the proposed solution to what biologists refer to as the sixth mass extinction (Kolbert 2014) is more of the violence through which this extinction has been made. In other words, while war may grant states and corporations access to land, it, and the enormous organization that constant war readiness demands, also contributes to the erosion of land. Yet, such resource shortage is also managed through martial means: through war or the threat of war.

Approaching this history and present development from an International Relations perspective, it can be argued that the need to secure access to Cheap Nature and the willingness to use violence to do so have been central to state and interstate relations since the early emergence of the capitalist world-ecology. The *geopolitics* of the capitalist world-system has thus been characterized by competition, conflict, and war since its inception. Discussing this precise history, political philosopher Jairus Grove (2019) has influentially claimed that contemporary extractive geopolitics has emerged out of 'an ecological principle of world making that renders some forms of life principle and other forms of life useful or inconsequential' (p. 3). The driver of this development, Grove notes, has been the capitalist project launched by a number of European nations and developed further by the US so that '[n]o anthropogenic, planetary-scale threat faced today—be it nuclear weapons, plastic, climate change, or global war—originated outside the Euro-American circuit of expansion, extractivism, and settlement' (p. 11). In this way, Grove identifies the emergence of a martial and extractive geopolitics as a main driver of what he calls extinction.

Another way of phrasing this is to say that an extractive geopolitics has steered the planet towards an epochal world-ecological crisis. So far, the US, as the hegemon of the world-system, has been spared much of the most appalling violence. However, now that the world-system's access to Cheap Nature is diminishing, this extractive geopolitics may accelerate and deepen international and national armed conflict to such an extent that the US may be drawn into very extensive armed conflict. Marzec is describing this possible development when he names the desperate attempt to manage the depletion of Cheap Nature a 'new geopolitics of diminishing resources' (Marzec 2015, p. 107). Similarly, if from a very different intellectual vantage point, International Relations scholar James E. Lee assumes in his book *Climate Change and Armed Conflict: Hot and Cold Wars* (2009) that the arrival of the 'Anthropogene Warming Period' (p. 25) will produce significant interstate and intrastate conflict:

> The Climate Change War will be a global period of instability that will last centuries. The period of greatest instability will be the twenty-first century. As in the Cold War, it will be a long struggle over core issues regarding rights and responsibilities in society. Throughout this period, there will be a new Cold War, and an existing Hot War that will intensify. Changes in climate will produce unique types and modes of conflict, redefine the value of important resources, and create new challenges to maintaining social order and stability. (p. 2)

While Lee's analysis builds on the idea that (a universalised) humanity has produced global warming, his prediction that biospheric erosion will cause tension and military conflict collates well with Grove's and Marzec's analyses. Of course, this development is already underway. International relations and peace and development research that examines the relationship between global warming and state or intrastate relations has revealed 'close links between climate drivers and armed conflict' in Africa (Regan and Kim 2020, p. 128) and Syria (Kelley et al. 2015). This work also tends to agree that 'it is increasingly becoming clear that climate change has already become a major factor in conflicts, and its influence is only going to increase over time' (Berhe 2022, p. 2). This can be perceived as the arrival of the 'new geopolitics of diminishing resources' mentioned by Marzec, but this geopolitics is 'new' only in the sense that it is practised at a time when the capitalist world-ecology has depleted Cheap Nature to such an extent that it is difficult to come by. As Grove and Marzec both contend, as Moore stresses, and as this book has repeatedly argued, the use of violence to establish commodity frontiers and to secure extraction has long been a part of the Euro-American hegemonic project.

Building on this understanding of what produces violence within the world-system, this chapter focuses on post-millennial American Climate Emergency Narratives that speculate on how the geopolitics of the existing world-system will seek to manage the ongoing socio-ecological breakdown and the depletion of Cheap Nature. The texts studied in the chapter thus take place in futures where unfolding world-ecological crisis has provoked the dominant nation-states of the world-system to go to war to ensure access to the energy, food sources, and labour relations that sustain them. The primary example is *Battlefield 2042* (ElectronicArts 2021)—a big-budget video game that involves the gamer in an ongoing global military conflict over dwindling resources and (thus) over world-system hegemony. This game thus speculates that the resource shortages and migration movement that biospheric erosion causes will produce an extensive 'climate change war' and the point of the game is to involve the gamer in such a war. This chapter moves on to consider American Climate Emergency Narratives where existing geopolitical relations have changed drastically and a new nation-state has risen to power to replace the US as the world hegemon. To this effect, the chapter discusses *Ship Breaker* (2011) and *The Drowning Cities* (2012), the first two novels of Paulo Bacigalupi's *Ship Breaker* trilogy. These novels play out in futures where global war has resulted in a profound transformation of the world-system. The capitalist world-system is still in place, but the US has become peripheral to its function in the sense that much of it is now a precarious and violent commodity frontier.

Ecological Erosion and the Future Battlefields of the World-System

That geopolitical conflict and war are central to much fiction set in futures transformed by biospheric erosion has been noted by scholarship. In particular, Ben De Bruyn's article 'The Hot War: Climate, Security, Fiction' (2018) describes how novels set in futures transformed by climate change can be seen as 'forecasting' climate war. De Bruyn's primary example is Tobias Buckell's techno-thriller *Arctic Rising* (2013) a novel that considers how global warming and an Artic free of ice open up new venues for exploitation and extraction, but also how this development creates significant geopolitical tension in the northern hemisphere and beyond. Other texts that predict the possibility of what Lee (2009) calls Climate Change War are Clive Cussler's *Artic Drift* (2009) and Matthew Glass's *Ultimatum* (2009). These novels describe 'global warming' as an important 'threat multiplier' (Gilbert 2012), but the wars that loom in the novels' background never break out. The skilled machinations of US science and military intelligence (Cussler) or US political diplomacy (Glass) are able to both cool down international relations and avert climate change. Thus, in these texts, a combination of traditional

Cold War practices and a collaborative geopolitics ushers the world into another revolution of now ecologically sustainable American world-system hegemony.

In other texts, Lee's (2009) 'Hot War' has become a reality. This war is either the explicit focus of the text or it is a historical fact that has drastically transformed the organization of the world-system. An especially relevant example is the video game *Battlefield 2042* (2021). This is a fiction from the core in much the same way as the major Hollywood films discussed in previous chapters. With a budget similar to that of a major Hollywood movie, the production process of *Battlefield 2042* involved a great number of game designers, coders, game testers, voice actors, and marketing firms located in the US but also in other parts of the world. The actual game was coded by the Stockholm studio DICE which launched the Battlefield franchise in 2002 with the release of the *Battlefield 1942*. Like all subsequent Battlefield games, this contained a multiplayer component where gamers collaborate as part of teams, fighting other teams to seize control of a virtual territory. To date, the franchise contains 12 major releases and several expansions for each release. The publisher of the Battlefield games is US-based Electronic Arts, one of the biggest digital game companies in the world.

Unlike many other electronic war games such as *Full Spectrum Warrior* (2004), or *Americas Army* (2002–2022), *Battlefield 2042* was not developed by the US DoD, and it did not receive funding from the US DoD or from any major manufacturer of hand-held weapons. That noted, like most war games, *Battlefield 2042* functions very much like the war games that have received such funding. Made primarily to cater to boys and young men who spend money and time performing virtual war on video consoles and PCs, the game puts the gamer into the body of a proficient soldier equipped with multiple weapons and then inserts this virtual body into intense military battles. The gamer views the world through the eyes of this soldier and can shoot at other soldiers with various guns held in the soldier's hands. Thus, this type of game is typically referred to as a (military) first-person shooter (FPS). The soldier avatar can (and will be) shot and blown up on several occasions during a gaming session, upon which it will respawn in a sheltered location so that the gamer can rejoin the battle.

Battlefield 2042 is one of many first-person shooters that take place in a future transformed by anthropogenic/capitalogenic violence. Other game franchises such as *Fallout* (1997–2018), *Stalker* (2007–2009), and *Metro*

(2010–2019) also explore worlds drastically transformed by atomic war or other violence done to the planet because of major military interstate conflict. In these series, the gamer takes on the role of a lone survivor who battles through the wasteland that such conflict has left in its wake. *Battlefield 2042* is different, firstly, because it takes place in a world where 'climate change', rather than atomic war, is the clearly stated reason for global conflict; and secondly because it inserts the subject into the conflict as such, rather than into its aftermath. An additional and crucial difference is that *Battlefield 2042* does not contain the story-driven, single-player campaign that is the centre of most other FPS games. Instead, the game tells its story through a set of game maps where a large number of players fight alongside each other, against other players or avatars (bots) operated by the game engine. Unlike most other first-person shooter games, this means that *Battlefield 2042* does not have a narrative ending where all objectives of a particular storyline have been reached. Instead, the gamer replays the same game maps, allowing individual gamers and teams to climb rankings much like athletes and sports teams do during a season in real life.[2]

The lack of a single-player campaign does not mean that there is no narrative context, however. Such context is provided by computer-animated 'cut-scenes': short video sequences that both set the mood of the game and describe the background of the military conflict that the gamer becomes part of during gameplay. The first such cut scene opens the first stage of the game. In this scene, the gamer is shown images of a cloud-covered planet and hears a gravelly voice saying: 'Fire, flint, language, machines - in the face of crises, humanity adapted. We became warriors, explorers, builders, dreamers flying forward. Take control of a changing world'. This very concise history describes the rise and evolution of humankind as a series of crises where humans have battled nature and other (less-than-human?) antagonists. According to this history, humanity was born out of its ability to control fire. This led to the construction of the first tools, then to language, then to complex machines. The same history also centres certain subjectivities very closely tied to the development of colonial capitalism: the soldier, the explorer, and the construction

---

[2] Like most (multiplayer) games today, *Battlefield 2042* constantly evolves with new game maps, bug fixes, and game features being added. Since its release in 2021, four major updates (called 'Seasons') with new downloadable content have been released.

worker. As such, this opening is in fundamental agreement with Will Steffen's et al. (2016) claim that the human species has been programmed to engineer the Anthropocene since the homo erectus figured out how to control fire a couple of million years ago.

This introductory statement is followed with a series of short and distorted news snippets that describe how extreme weather events (including 'the world's first category 6 storm'), rising sea levels, and uncontrollable forest fires destroy ecologies vital to human existence (Fig. 5.1).

Other short news items outline the arrival of mass migration, a new global Great Depression, and violent social unrest across the world. Most African, but also many European nations are described as breaking down almost entirely. In the wake of this development, two major powers are said to remain: the US and Russia. These are challenged by a disorganized 1 billion large refugee community that refers to itself as the Non-Patriated or No-Pats. This sets the stage for the global war for resource control that is the core of the game. When the cutscene ends, the gamer is invited to perform precisely such a war.

Russia has been part of several Battlefield games and other military multiplayer franchises. Since the Cold War, this nation has often been cast as a major international player and a constant looming threat to continued US world-system hegemony. It is thus not surprising that Russia features in the game as the (playable) extractive thug who refuses to let go of the fossil-fuel economy. What is surprising, however, is the complete absence of China. When the opening cut scene narrates the history of biospheric collapse and world-system transformation, there is virtually no mention of China. The reason for this is most likely the fact that China has objected to being depicted as a belligerent nation and a participant in future wars for world-system hegemony. In 2010, director Dan Bradley completed his film *Red Dawn* (2012), a remake of the Hollywood blockbuster from 1984 by the same name. Like the original film, Bradley's updated version described a massive military invasion of US soil, but he replaced the Soviet forces of the original with a Chinese enemy. In view of the growing importance of the Chinese box office, and because Chinese officials protested, the film was reshot and digitally altered to instead show a (now united and capitalist) Korea as the invader. Also, a year after *Red Dawn* had premiered, *Battlefield 4* (2013) was banned from the Chinese market because it portrayed China as one of the playable superpowers in the future world war it portrays (Valeriano and Habel 2016). It is thus not

Fig. 5.1 Screenshots from the opening cutscene of *Battlefield 2042*

surprising that China does not feature in *Battlefield 2042*. At the same time, it is possible to think of the Russian forces in *Battlefield 2042* as a kind of ersatz China; a stand-in for this nation similar to North Korea in *Red Dawn*.[3]

---

[3] *Red Dawn* was turned into a computer game: *Homefront* (2011) that also features a Korean invasion of the US.

The background story of ecological emergency and social unrest clearly builds on the vision of the future that Lee (2009) and Grove (2019) predict in different ways and from different perspectives. In *Battlefield 2042*, an eroding biosphere, resource shortage, and consequent capitalist crisis are what have caused the war that the gamer participates in. War is thus portrayed as a kind of inevitable geopolitical adaptation to the resource scarcity experienced by capitalism in a time of socio-ecological breakdown. In fact, the heroic, underdog No-Pats describe themselves as people who are precisely *adapting* through war. In the introductory cut scene, the narrator exclaims, to building, triumphant music, that: 'We are the warriors, we are the adaptation, and this fight…is ours'. As the actual gameplay makes clear, adaptation means adjusting to both the pressures of a changing geopolitical landscape where the two remaining superpowers vie for control, and to the fast violence produced by a damaged biospheric system. Thus, the gamer must overcome a hostile military force, but also negotiate and overcome the various extreme weather events that are apparently a regular occurrence in this transformed socio-ecological future.

This becomes evident when the gamer enters the opening game map set in Doha, the capital city of Qatar. When first joining the unfolding battle, the sun is shining, making it easy to locate and fire on enemies using one of the many weapons at the gamer's disposal. However, before long the wind picks up and a gigantic sandstorm enters the city, reducing visibility and enveloping the entire city in a red haze (Fig. 5.2).

Similar extreme weather events occur on many other game maps. In 'Orbital', set at a satellite launching facility in French Guiana, the gamer is faced not only by opposing US or Russian forces (depending on which of these nations states the gamer has decided to join before the battle begins), but also by an enormous tornado that sucks people, vehicles, and other moveable objects up into a dark sky (Fig. 5.3). The game map 'Kaleidoscope' which takes place in Songdo, South Korea, also features a tornado, accompanied by thunder, lightning, and torrential rain.

In addition to featuring extreme weather, the maps narrate future socio-ecological breakdown through the design of the virtual worlds the gamer's avatar traverses. What this means is that the maps make the effects that continued socio-ecological breakdown is likely to have on natural and urban environments and capitalist infrastructure visible. The game map 'Breakaway' takes place on a 'partially thawed plateau in Antarctica, the site of [Russian] illegal oil drilling and a petroleum refinery'.

Fig. 5.2 Sandstorm on its way towards the city of Doha in *Battlefield 2042*

Fig. 5.3 A tornado complicates the military battle on the game map 'Orbital' in *Battlefield 2042*

'Renewal' is set in Egypt where half the map is reclaimed desert turned to lush, green plantations and the other half consists of solar farms, and in 'Stranded', the gamer fights around a supertanker stranded in a now dried-out Panama Canal (Fig. 5.4).

Fig. 5.4 The game map 'Stranded' showing a beached supertanker and a dried-out Panama Canal in *Battlefield 2042*

To return to the game's introductory cutscene, *Battlefield 2042* proposes that biospheric erosion is anthropogenic. There are frequent references to a universal 'us' and the general idea is that the ecological disaster that forms the game's backstory and rationale, and that also informs both the geography and the gameplay of the various maps, has been caused by 'humanity'. That said, the game maps still tacitly register extractive capitalism as an engine of socio-ecological breakdown. The stranded tanker in the Panama Canal, a ship-breaking yard on the Indian west coast, and the Russian oil wells and petroleum refineries in Antarctica are not simply in the game as colourful milieus within which war is fought, they also register the reasons *why* the biosphere has collapsed. These game maps insert the gamer into the debris of a transformed, combative—yet still functional—capitalist world-system, and by doing so, they gesture to this system's culpability in the collapse of the biosphere. The stranded supercarrier, the enormous ship-breaking yard, and the still-operational oil extraction facilities in what used to be a protected Antarctic ecology are most usefully read as physical/virtual manifestations of the extractive processes that have brought on the war that the gamer participates in. In addition to this, the No-Pats into whose shoes the gamer frequently steps are described by *Battlefield 2042* as a rebellious, anti-state, and anti-extractive group. Thus, the No-Pats are defined as 'the once privileged and the impoverished with backgrounds that are worlds apart, forced together, determined to survive'. They are furthermore 'distrustful of the governments that exiled them' and they 'refuse calls to reassimilate'. This also suggests a radical stance; a willingness to transform the violent and extractive system that is causing the Earth System to erode in the present.

But as the game progresses, it becomes clear that *Battlefield 2042* is not a game that encourages revolution and system change in any sense of the word. Although the background story, the game maps and the avatars at the gamer's disposal register a certain awareness of the role that extractive capitalism has played for climate change, the actual gameplay forces the gamer to take part in a fundamentally adaptive and enormously violent Climate Change War. As described, at the start of each multi-player game, the gamer takes on the role of one of two sides, both of which are tasked with securing a particular objective. Before entering the multiplayer map 'Breakaway', set in Antarctica, the gamer is briefed either as a member of the offensive, US contingent or as the defending Russian side. If on the side of the American force, the gamer is told that 'This is it, people. It's time to put a stop to the Russians' illegal Antarctic oil drilling. The first step will be removing any Russian military presence from the region'. If the gamer plays as part of the Russian defenders, the briefing instead reads: 'We have an *emergency*: The US has deployed forces against our Antarctic oil drilling operation. That facility is critical to our energy infrastructure, and we must defend it at all costs. Stop the Americans from securing the sectors' (my italics). In both cases, as Marzec (2015) puts it, the 'environment is made visible in terms of its ability to yield energy' (p. 106). When the environment is perceived through this perspective, it becomes, as Marzec has observed, a military concern. Each campaign revolves around the imminent need to secure a particular area within the game space, of adapting to the challenges that enemy soldiers and the hyper-charged weather present the gamer with, and to thus play an active, if imaginary and virtual, military geopolitical role at a time of world-system confrontation caused by socio-ecological breakdown.

This fundamentally conservative, core understanding of the biosphere as a repository of (cheap) extractable energy is further brought out by the gameplay. Game theorist Ian Bogost (2007) has influentially suggested that games produce ideology not simply via their visuals and (written) textual content, but also through the process of gameplay itself. Bogost terms this ludic quality *procedural rhetoric* and argues that games persuade primarily through 'rule-based representations and interactions' (p. ix).

Thus, electronic games are not just visual and textual objects, but, importantly 'computational artifacts that have cultural meaning *as* artifacts' (p. ix, italics in the original). The procedural rhetoric unfolding in games can work to 'support existing social and cultural positions', but it can also 'disrupt and change fundamental attitudes and beliefs about the world' (p. ix). In the case of *Battlefield 2042*, the visual and textual surface layer of the game can be said to include elements that critique the socioecological violence performed by extractive capitalism. However, on the procedural level, the game imagines world-system geopolitics as the only viable paradigm. Once the gamer's heavily armed avatar runs through the streets of Doha, or negotiates a barren Antarctica, resolving a clearly defined battle objective, the ongoing socio-ecological breakdown is effectively reduced to a geopolitical resource crisis and thus, it is accessible and comprehensible only in the form of an engagement opportunity.

In other words, *Battlefield 2042* can be defined as a procedural Climate Emergency Narrative from the core that inserts the gamer into a 'new geopolitics of diminishing resources' (p. 107). As with most of the previously discussed narratives of this book, the violent world-system geopolitics that informs the game—the type of capitalist, life-making it envisions—understands the environment not as a multispecies, planetary habitat into which humans are folded, but as a landscape of dissipating yet vital energy resources. As an effect of this understanding, the environment is produced as a stage for a series of endlessly repeated global, military campaigns the object of which is to secure territory. While the game caters to an international audience and predicts that the existing social and economic order will produce catastrophic biospheric erosion, it never suggests that there is an alternative to the world-system that brought it on, or to the military violence it involves the gamer in. Also, while it imagines an uneven conflict between three contenders: America, Russia, and the No-Pats, and forges an allegiance with the No-Pats who have been deprived of national belonging, the game never moves to strip America of its status as hegemon of the world-system. Once the action begins, the game's procedural rhetoric is solely focused on the rehearsal of spectacular violence that ultimately aims to preserve the world-system as such.

At the same time, the game's procedural rhetoric also connects with Grove's contention that a geopolitics of war is not simply a way for nation-states to secure access to land and resources, but a 'form of life' (p. 3). As such, this geopolitics is an 'embodied becoming [...] written into the very musculature of our bodies, practices, and communities' (p. 6). The military war game involves its (predominantly male) audience in a virtual version of such becoming. By definition, a war game does not allow for any solution to the geopolitical challenges they envision except war. This is the point of the war game. If war is one of several possible geopolitical strategies in the real world, it is, in such games, the only strategy. This, in turn, makes war the only form of life that can be imagined and enacted while playing the game. In addition to this, the physical, muscular commitment of the gamer—the act of moving certain muscles and engaging the brain to beat opposing gamers in battle—is also noteworthy. While it is no surprise that war is written into the bodies of professional soldiers, it is striking that this occurs, if in a slightly different way, also for the millions of civilians who practice FPS war gaming. Becoming competitive in *Battlefield 2049* or any other military multiplayer game demands constant practice and the development of certain muscles and muscle memories that enable very fast reactions. In this way, the military shooter can be described as a performative text where a geopolitics of war is written into minds, bodies, practices, and communities.

NEW WORLD-SYSTEM HEGEMONS

If *Battlefield 2042* takes place in the midst of a future Climate Change War, much of the fiction produced by trend-setting 'climate fiction' author Paolo Bacigalupi is located in more distant futures where socio-ecological breakdown has run its course. In these futures, the world-system is still very present and keeps organizing the social, economic, military, and extractive order, but the nation-state of America has become peripheral to a new core. The notion that socio-ecological breakdown may lead to a reshuffling of the world-system itself rather than, or as a preliminary to, the demise of the world-system as such, has been an important topic to Political Science/Philosophy, International Relations, and world-system scholarship. In *Savage Ecology*, Grove raises the question if America will remain the hegemon 'long enough for China to put its stamp on the human apocalypse' (p. 23). Grove thus argues that while 'we' currently 'live in the death rattle of *Pax Americana*' (p. 23, italics in the original), he also speculates that the next stage of terminal world-system crisis may not be the demise of capitalism as a world order, but rather 'the self-destruction of the United States' (p. 23) to the benefit of China, a nation that would then ascend to the position of core hegemon. Such a transformation of the organization of the world-system would not change the nature of extractive, militant geopolitics—people would still be living in what Grove describes as 'the shadow of an annihilating repetition' (p. 278)—but this process would be steered by China rather than by the Euro-American military-industrial complex. Again, in the wake of such a transformation, America might well be transformed into a commodity periphery servicing a reorganized world-system: the target of the same extractive and violent geopolitics the nation once established.

The possibility that the American era might be over and that China will become the new hegemon of the world-system is also considered by influential International Relations scholar John Ikenberry, a firm believer in the Euro-American 'liberal project'. To Ikenberry, the questions are whether an increasingly powerful China will 'overthrow the existing [liberal and democratic] order or become a part of it' (2008, p. 23), and if the US can do anything to maintain its dominant position as China's influence grows. In *Adam Smith in Beijing: Lineages of the 21st Century* (2009), Giovanni Arrighi discusses the same development from a world-systems perspective that does not assume that US liberal capitalism is the ultimate global social order. To Arrighi, the rise of China is not so much a question of what type of politics (democratic, autocratic, liberal) will structure the future capitalist world-system, as evidence that the US is losing its hegemonic position. As Arrighi shows, and as other world-system research also reveals, the US economy has been faltering since the 1970s and it seems likely that the nation 'will continue to decline' (Chase-Dunn et al. 2005, p. 233), a development that may open the door for Chinese hegemony. Again, such a transition does not change the fact that the world-ecology is moving towards an epochal crisis. Rather, as Moore (2015) has argued, 'the rise of China' (p. 86) is a direct consequence of extractive capitalism's pursuit of the Four Cheaps (labour power, food, energy, and raw material). In this way, China's rise can be considered as a stage of the wider epochal crisis, as well as an acceleration of this crisis.

Bacigalupi's *Ship Breaker* trilogy takes place in a future where the composition of the world-system has been fundamentally altered, but where capitalism has transitioned away from the ecocidal fossil-fuel economy. *Ship Breaker* (2010), the first novel of the trilogy, opens on a stretch of coast close to what used to be New Orleans. The protagonist is the teenage boy Nailer who makes a precarious day-by-day living by stripping now antiquated and stranded ships of their electronics and cables.

Importantly, and as I have discussed elsewhere (Höglund 2020), this is a setting borrowed from the Global South and existing ship-breaking yards such as Chittagong in Bangladesh.[4] The material that Nailer and the other members of his crew scavenge is then sold on to the militarized mega corporations that have become central to how the new geopolitical order that looms in the background operates. While this refashioned world-system has turned to more sustainable energy forms (this is a world-system accelerated by Green capitalism), it remains fundamentally unequal and unjust. Thus, the life that Nailer leads is profoundly precarious and unsustainable. The damage already done to the biosphere produces enormous hurricanes that regularly pommel the shanty town that he and other workers live in. Also, Nailer knows that he is growing too big for the light crew he is on, and even if he survives the dangerous work of mining asbestos-lined ducts of copper wire to move on to some other job, his criminal, sadistic, and drug-addicted father is likely to either kill him or force him to join his gang. He must exit this dangerous place and find another and less precarious place to call home.

The second novel of the trilogy, *The Drowned Cities* (2012), focuses on the 'war maggot' Mahlia. In her early teens, she lives a miserable life in a poor and exposed village in what used to be the US Southeast, a few days' journey from the Washington urban sprawl that has now become the titular drowned cities. As in *Ship Breaker*, it is clear from the context that global warming has inundated most major American coastal cities and turned much of the Southeast into a tropical swampland. The reader also surmises that the lack of resources and the turmoil produced by war and global socio-ecological breakdown must have eroded the US politically, financially, and militarily to such an extent that the nation-state has become defunct. Indeed, the world that Mahlia inhabits also looks strikingly like some of the most poor and war-ravaged parts of the Global South today. Warlords rule the land with the help of gangs consisting of drug-addled and traumatized child soldiers. The few survivors that remain have to endure the roaming gang's senseless violence, the fierce climate, and the pandemics that regularly visit the exposed communities.

Malia is especially vulnerable because she is the castoff daughter of an officer who was part of a now-abandoned Chinese peacekeeping effort designed to maintain order in this forlorn corner of the world-system. In

---

[4] The ship-breaking yard of the opening of the novel is also reminiscent of the previously discussed *Battlefield 2042* map 'Discarded', set on India's West Coast.

this way, the Ship Breaker trilogy clearly registers the prevalent fear that China might replace the US as the hegemon of the world-system. The fact that China has entered former US territory not as a military invader, but as a peacekeeper is telling. The former hegemon of the world-system has fallen so far from its former glory that it has become a new and tortured Global South and, as such, it is also the reluctant recipient of international (military) aid work. Mahlia's blood connection with the Chinese does not do her any favours. The roaming gangs of child soldiers view her as a traitor and collaborator. She has already lost one hand to mindless violence and survives in the present only because she assists the village doctor. Her luck will not hold. Like Nailer, she must move on.

Fortunately, both youngsters have something to barter. Mahlia remembers the life she had in what was once Washington before the Chinese peacekeepers left, leaving Mahlia and her mother to fend for themselves. She recalls the violence, the 'mobs and soldiers', and 'dripping machetes' (Bacigalupi 2012 p. 259), but she also remembers that her mother had acquired and hidden the 'treasure trove of a dead nation' (Bacigalupi 2012 p. 427) in their old rooms in the city. In his part of the defunct nation, Nailer has come across the young and beautiful Nita Patel. She has been shipwrecked close to the shipyard during a hurricane and, because she is a member of one of the most affluent entrepreneurial families in this new world, she can be his ticket out of his miserable existence. Encouraged by these resources, Nailer and Mahlia begin arduous treks through a landscape made hostile by both biospheric transformation and utter social collapse, towards the fortified borders that separate them from the new hegemons of the transformed world-system.

Nailer and Mahlia know very little about the 'dead nation' that used to rule the dangerous world they traverse. What used to be the US exists for them only in the shape of inundated ruins. That noted, Nailer and Mahlia also know the US through the still ongoing violent and extractive geopolitics that once propelled the US to the position of world hegemon. Indeed, the world-system and the geopolitical paradigm that the nation forged on top of the efforts of European imperialism are very much alive. It is this very paradigm that is turning people close to Nailer and Mahlia into addicts, into cheap, enslaved labour, or into child soldiers who eke out an enormously precarious existence in what has become a poorly maintained and war-ravaged commodity periphery. The only difference between this new world-ecology and the anglophone empire that came before it is that the new powerhouses of the system are far less reliant on fossil fuel and, of course, that the seats of power are not in Washington, D.C. or New York, but in Delhi, Beijing, and Tokyo.

It is into these new cores that Nailer and Mahlia hope to escape. They know such privileged worlds exist. They have seen the arrival and departure of white, wind- and solar-powered 'clipper ships' (Bacigalupi 2011, p. 7) that pick up the material extracted from the shipyard or an increasingly dismantled Washington Capitol. Again, the cores and privileged semiperipheries of this reorganized world-system are now located in Asia and governed by nation-states such as India, Japan (referred to as Nippon), and China, and run by international, militarized mega corporations such as the one operated by Nita's family. These control the still fully functional world-system alongside what remains of the world's semiperipheral nation-states.

In this way, Bacigalupi's trilogy casts the socio-ecological crisis as the end of American hegemony, but not of the capitalist world-system. The important point here is that this imaginary shift is meant to trouble the reader from the core. This is what your world will come to, novels such as these are telling their readers, if you continue down the petro-fuelled, ecocidal path you are currently on. Oceans will rise, cities will crumble, governments fall, new hegemons rise, and your sons and daughters will be scurrying in the ruins like the children in Asia and Africa whose unpaid labour you rely on in the present. This shift of power within the existing global social order is the emergency. This may gesture towards a critique of the ongoing militant and fundamentally extractive geopolitics that bind the current world order together and through which the periphery and semiperipheries are made to service the core. Yet, like many of the other texts discussed in this book, it is also a narrative that struggles to imagine worlds other than those that brought the crisis on. Mahlia and Nailer do not move on to forge new social worlds. Their exits from the peripheries into which they were born, across the border that separates them from the new geographies of power, merely serve to insert them into another location of the world-system. Again, the lesson learned is that borders must be fortified, crises prepared for, emergencies managed, and potential enemies pre-emptively combatted.

The transformation of the US into a dreadful commodity periphery is a common trope in the American Climate Emergency Narrative. Alongside Bacigalupi's fiction, Marcel Theroux's *Far North* (2009), and Mindy McGinnis' *Not a Drop to Drink* (2013) and *In a Handful of Dust* (2014) also move the story into futures where (white) protagonists live precarious lives outside of the protective embrace of US petromodernity. In *Far North*, the protagonist belongs to a now-defunct religious community that once escaped to Siberia, but, as in Bacigalupi's story, the system that kept this community operational has now collapsed and the land and the people on it are forced into slavery and made to scavenge the irradiated and poisoned ruins of the cities left behind. McGinnis's two novels are set in an arid part of the US where isolated households protect precious water sources from violent and/or plague-ridden wanderers. The protagonists of these novels are proficient killers, shaped by the need to defend the scant resources that their worlds can supply, but the novels also suggest that the world-system that produced the ecological crisis that is at the heart of the story still exists. Life in the periphery is difficult not only because it is dry and depleted, but because it is precisely a periphery within the remit of the new world-system.

Imagining Future War

Battlefield 2042 realistically proposes that what it describes as the climate emergency will produce, and will thus be experienced as, war. In other words, while socio-ecological breakdown may be directly experienced by some as flooding, forest fires, accelerated storms, and a shortage of energy, drinkable water, and food crops that demand stable conditions to grow, and by collapsing labour relations, it will also be encountered in the form of armed international conflict. Again, if what this text imagines as a climate crisis is ultimately an epochal crisis for the capitalist world-ecology, the unravelling of this crisis is likely to manifest through the economic and military intrastate and interstate conflict that is designed to keep this world-ecology running, to maintain US core hegemony, or, in the case of other nation-state actors, to challenge American dominance. Between the coded lines, *Battlefield 2042* registers that geopolitics is, as proposed by Grove, an instrument of petro-energized, capitalist world-making. In this way, the game tacitly observes that a long history of extractive geopolitics has eroded the Earth System. On a planet where resources are finite, the 'geopolitics of diminishing resources' (Marzec 2015, p. 107) is an inevitable successor both to neoliberal globalization and to the geopolitics of violent extractive expansion that preceded Earth-System erosion and the depletion of (cheap) resources. This geopolitics of diminishing resources is, in this game, ultimately a geopolitics of war.

In all the narratives discussed, the climate emergency is thus also a geopolitical security emergency that erupts into armed conflict. In *Battlefield 2042* and the *Ship Breaker* trilogy, human worlds have collapsed more due to the (geo)political developments that follow in the wake of biospheric erosion, than because of this erosion as such. In addition to this, in *Battlefield 2042*, the gamer *performs* future socio-ecological breakdown as precisely a military geopolitical security emergency. Indeed, *Battlefield 2042* inserts its audience, the gamer, into a world where such conflict is perpetual. *Battlefield 2042* is, like other multiplayer first-person military shooters, a game without an ending. The gamer is stuck rehearsing a procedural rhetoric that may recognize that war is extractive, but that still insists that there is no other recourse. Ultimately, the only story *Battlefield 2042* tells about socio-ecological breakdown is that war is inevitable. And also, a joy of sorts when it is experienced through a body in front of the computer or gaming console.

In this way, the American Climate Emergency Narrative represents socio-ecological breakdown as a geopolitical crisis that threatens to curb US access to easily extractable natural resources such as potable water, arable land, metals, and energies such as coal and oil. In the texts studied, this crisis has produced world-scale military conflict. While all texts discussed in this chapter can be said to register the possibility of epochal capitalist world-ecological crisis in this way, none of them envisions this crisis as terminal. *Ship Breaker* and *The Drowning Cities* locate the reader in worlds where the US has been unable to sustain its dominant position in the world-system. New powerhouses have risen to take the place once assumed by the US. In Bacigalupi's writing, China appears to be the new hegemon of a still thriving, if less ecocidal, capitalist world-ecology. The people who now inhabit what used to be the US are thus effectively confined within one of the new commodity peripheries of the reorganized world-system.

By adding the precarious No-Pats community to its static narrative, *Battlefield 2042* gestures vaguely towards the commodity periphery and the experience of being peripheralized and extracted. That said, as soon as the fighting starts, the gamer has all the agency in the world. Instead of being a pawn in the commodity periphery, the gamer becomes a soldier with a mission; not a subject who is secured but one who secures territory. The protagonists of *Ship Breaker* and *The Drowning Cities* have far less agency. Maimed and depleted, they suffer the precise hardships experienced in the most troubled parts of the world-system. They have no choice but to escape. In this way, these texts model the future of socio-ecological breakdown in the Global North on the inequalities that exist in the capitalist world-system today. When fiction from the core is made to enter futures beyond the current moment of pending epochal crisis, it seeks its dark mirror image within the worlds it has already depleted. Yet, when doing so, these texts rehearse the notion that no other world beyond the violent life-making capitalism engages in can take form when existing walls collapse.

Such possible futures exist, of course. Deeply aware of how geopolitics in the international system has been geared towards war and conflict, Simon Dalby (2007) has promoted the emergence of a new and radically different 'Anthropocene geopolitics'. Dalby describes this as a collaborative geopolitics designed to alleviate the tensions and conflicts that are both generating human death and suffering and preventing powerful nation-states and nation-state alliances from reaching agreements on how to limit the release of greenhouse gases into the biosphere. The question is if the introduction of such an Anthropocene geopolitics ultimately serves to salvage the capitalist world-system, by making it more sustainable. Grove (2019) is far less optimistic. If the 'geopolitical project of planet Earth is a violent pursuit of a form of life at the cost of others' (p. 3), geopolitics can never function as a foundation for meaningful social and ecological change. Something very different is needed. Thus, Grove does not propose to alter geopolitics. Instead, he wishes for a future social and ecological order that emerges out of 'a social sciences for Earthlings' and that rely on 'minor traditions, incipient practices, novel senses of belonging, and anachronistic forms of life, both futural and deeply old' (pp. 278–279). Moore's alternative is similar but much more clearly political and organized than that envisioned by Grove. If capitalism valued 'labor productivity organized through the exploitation of labor-power and the appropriation of Cheap Nature', a 'sustainable and socialist law of value would privilege the healthy, equitable, and democratic relations of reproduction for *all* nature' (2015, p. 294). If such a law of value were to inform international relations, a fundamentally different geopolitics—worthy of a different name—is perhaps possible. Some of the texts discussed in Chapter 8 investigate precisely such a prospect.

WORKS CITED

Arrighi, Giovanni. 2009. *Adam Smith in Beijing: Lineages of the 21st Century.* London: Verso.

Bacigalupi, Paolo. 2011. *Ship Breaker.* London: Atom.

———. 2012. *The Drowned Cities.* London: Atom.

Berhe, Asmeret Asefaw. 2022. 'On the Relationship of Armed Conflicts with Climate Change.' *PLOS Climate* 1 (6): e0000038.

Bogost, Ian. 2007. *Persuasive Games: The Expressive Power of Videogames.* Cambridge, MA: MIT Press.

Bonneuil, Christophe, and Jean-Baptiste Fressoz. 2016. *The Shock of the Anthropocene: The Earth, History and Us.* London: Verso.

Buckell, Tobias S. 2013. *Arctic Rising.* London: Random House.

Chase-Dunn, Christopher, Andrew K. Jorgenson, Thomas E. Reifer, and Shoon Lio. 2005. 'The Trajectory of the United States in the World-System: A Quantitative Reflection.' *Sociological Perspectives* 48 (2): 233–254.

Cussler, Clive, and Dirk Cussler. 2009. *Arctic Drift.* London: Penguin.

Dalby, Simon. 2007. 'Anthropocene Geopolitics: Globalisation, Empire, Environment and Critique.' *Geography Compass* 1 (1): 103–118.

De Bruyn, Ben. 2018. 'The Hot War: Climate, Security, Fiction.' *Studies in the Novel* 50 (1): 43–67.

Electronic Arts. 2021. *Battlefield 2042.*

Gilbert, Emily. 2012. 'The Militarization of Climate Change.' *ACME: An International Journal for Critical Geographies* 11 (1): 1–14.

Glass, Matthew. 2009. *Ultimatum.* New York: Grove Atlantic.

Grove, Jairus Victor. 2019. *Savage Ecology: War and Geopolitics at the End of the World.* Durham: Duke University Press.

Höglund, Johan. 2020. 'Challenging Ecoprecarity in Paolo Bacigalupi's Ship Breaker Trilogy.' *Journal of Postcolonial Writing* 56 (4): 447–459.

Ikenberry, G. John. 2008. 'The Rise of China and the Future of the West-Can the Liberal System Survive.' *Foreign Affairs* 87 (1): 23–37.

Kelley, Colin P., Shahrzad Mohtadi, Mark A. Cane, Richard Seager, and Yochanan Kushnir. 2015. 'Climate Change in the Fertile Crescent and Implications of the Recent Syrian Drought.' *Proceedings of the National Academy of Sciences* 112 (11): 3241–3246.

Kolbert, Elizabeth. 2014. *The Sixth Extinction: An Unnatural History.* New York: Henry Holt.

Lee, James R. 2009. *Climate Change and Armed Conflict: Hot and Cold Wars.* London: Routledge.

Marzec, Robert P. 2015. *Militarizing the Environment: Climate Change and the Security State.* Minneapolis: University of Minnesota Press.

McBrien, Justin. 2016. 'Accumulating Extinction: Planetary Catastrophism in the Necrocene.' In *Anthropocene or Capitalocene? Nature, History, and the Crisis of Capitalism*, edited by Jason W. Moore, 116–137. Oakland: PM Press.

McGinnis, Mindy. 2013. *Not a Drop to Drink*. New York: Katherine Tegen Books.

———. 2014. *In a Handful of Dust*. New York: Katherine Tegen Books.

Moore, Jason W. 2015. *Capitalism in the Web of Life: Ecology and the Accumulation of Capital*. London: Verso.

Regan, Patrick M, and Hyun Kim. 2020. 'Water Scarcity, Climate Adaptation, and Armed Conflict: Insights from Africa.' *Regional Environmental Change* 20: 1–14.

Steffen, Will, Paul J. Crutzen, and John R. McNeill. 2016. 'The Anthropocene: Are Humans Now Overwhelming the Great Forces of Nature?' *Ambio* 36 (8): 614–621.

Theroux, Marcel. 2009. *Far North*. London: Faber & Faber.

Valeriano, Brandon, and Philip Habel. 2016. 'Who Are the Enemies? The Visual Framing of Enemies in Digital Games.' *International Studies Review* 18 (3): 462–486.

CHAPTER 6

The Displaced

DISPLACEMENT, REFUGE, AND RUPTURE

Socio-ecological breakdown and displacement go hand in hand. As described in the first chapters of this book, the violence employed by agents of the capitalist world-system to clear the land of Indigenous people and species forced survivors to escape and seek refuge elsewhere. The extractive practices that followed and that sought to source nature on the cheap initiated a global ecological breakdown that soon took the form of soil erosion and contamination, and that has now evolved further into hyper-charged storms, droughts, and floods, and slowly increasing global warming (Seneviratne et al. 2012; Ummenhofer and Meehl 2017; Padrón et al. 2020). This erosion of the Earth System is now producing new exoduses, as people must abandon homes and lands that cannot feed or shield them any longer. As Xu et al. (2020) have argued, the Holocene 'environmental niche' that humans have taken advantage of during the past millennia, where the mean temperature is around 13 °C, is slowly disappearing in many places and being replaced by hotter zones. By the year 2070, Xu et al. estimate, '1 to 3 billion people' (p. 11350) will live in parts of the world where the climate niche that sustains human life today no longer exists. These people will have to abandon these regions, or slowly succumb to heat death and starvation. The problem is that there are few places to escape to. If the Holocene was a time when species could find refuge and recuperate, the Capitalocene is a period when such

© The Author(s) 2024   131
J. Höglund, *The American Climate Emergency Narrative*,
New Comparisons in World Literature,
https://doi.org/10.1007/978-3-031-60645-8_6

refuges are rapidly disappearing. As Donna Haraway (2016) has observed with Anna Tsing, 'the earth is full of refugees, human and not, without refuge' (p. 100).

This search for refuge is likely to contribute to the violent geopolitical conflicts and depleted futures discussed in the previous chapter. If military conflict increases, it will greatly accelerate the displacement caused by a warming planet (and its underlying causes). A briefing commissioned by the European Parliament estimates that in 2020 alone, more than 30 million people were displaced by environmental disasters linked to climate change (Apap and Revel 2021). In the coming 50 years, as the biosphere continues to warm and as capitalism intensifies its pursuit of Cheap Nature, this number will most likely increase drastically. Unless measures are taken to significantly mitigate socio-ecological breakdown, it is likely that between 200 million (Wyett 2014) to 1 billion (IEP 2020), will find themselves displaced by the unravelling of the capitalist world-ecology by the year 2050.

In this way, socio-ecological breakdown and the armed conflicts meant to secure the extractive potential that remains in various parts of the world, are already creating a slow but inexorable movement of people from the most affected areas into parts of the world where conditions for human and extra-human life are still decent, or where the economic resources that create social resilience are present. Such areas are obviously located mostly in the core and the privileged semiperipheries of the world-system, but as the number of refugees surges, governments in these parts of the world-system are increasingly reluctant to provide necessary sanctuary. As I will return to in this chapter, politicians in privileged parts of the world are now making grand careers out of the promise to keep refugees out of their nation-states.

With these developments in mind, this chapter moves on from the large-scale geopolitical confrontations discussed in the previous chapter, to instead centre the unsanctioned but inevitable human and extra-human mobilities that the interconnected and cascading erosion of social, economic, ecological, and atmospheric worlds produces. The chapter thus shows how the American Climate Emergency Narrative imagines the relationships between displacement, mobility, the biospheric crisis, the depletion of Cheap Nature, and the systems designed to keep capitalist modernity operational. In studying this development, I am building on already existing literary scholarship. Bryan Yazell (2020) has importantly observed that the climate migrant or refugee displaced in a world destabilized by biospheric erosion is a common figure in what he terms climate fiction. Similarly, Ben De Bruyn (2020) has investigated a set of British literary novels that centre climate-induced migration, arguing that these texts help readers to move 'beyond [the] simplistic militarized and humanitarian frames' (p. 1) offered by anti-migration populist discourse. Adding to this work, this chapter discusses how the American Climate Emergency Narrative, as a type of fiction produced in an increasingly fortified hegemonic core of the world-system, portrays displacement. The chapter focuses on two central tropes that structure this narrative. The first such trope appears in texts that cast the racialized climate migrant or refugee as a border-scaling and monstrous being that produces an acute state of insecurity. In this conservative and often perversely violent narrative, exemplified by Marc Forster's film *World War Z* (2013), the only way forward is to make full use of the vast military and scientific arsenal that the core controls. The second trope centres on white Americans who have been deprived of land, livelihood, and dignity, and who are forced to enter fundamentally insecure ecological and political spaces. This part of the chapter briefly considers Cormac McCarthy's *The Road* (2006), but it primarily discusses Brian Hart's *Trouble No Man* (2019). As the chapter argues, narratives such as these novels are instructive in the sense that they help readers in the Global North to understand what the consequences might be for people of the core if nothing is done to prevent further biospheric erosion. Even so, and as a type of Climate Emergency Narrative, these texts centre on the future plight of white, middle-class protagonists and fall short of imagining a world beyond borders, border violence, and border thinking.

MOBILITIES, BORDERS, AND THE MIGRANT

The official descriptor for those fleeing land that has been depleted by the pursuit of Cheap Nature and by the erosion of the climate is 'climate migrant'. According to the definition, 'climate migrants' are people who have chosen, rather than having been forced, to leave their homelands. As *migrants*, rather than *refugees*, they are not given refugee status and thus do not have the right to international protection defined by the UN in the *Convention Relating to the Status of Refugees* passed in 1951. When crossing national borders as migrants rather than as refugees, they can be, and routinely are, deported back into the hostile environments they escaped from.[1] In this way, climate displacement is perceived, by the security institutions of the Global North, as a security crisis. This crisis is typically mediated with the help of the established national security paradigm (Vietti and Scribner 2013; d'Appollonia 2017) that casts international and internal migration movements as putting unsustainable stress on border maintenance, federal and state policing efforts, the National Guard, the social and health security apparatus, and the economy of the nation-state generally.

In addition to this, corporations central to the operation of the extractive capitalist world-ecology perceive climate migration as problematic because it creates a shortage of cheap labour at sites of extraction in the Global South and in the semiperipheral areas of the US. Todd Miller shows in *Storming the Wall: Climate Change, Migration, and Homeland Security* (2017), how this development is another reason why the global security apparatus worries about socio-ecological breakdown. Miller thus reveals how players within the military-industrial complex are aware of, and deeply concerned by, the fact that socio-ecological erosion is forcing people in places such as Sub-Saharan Africa to abandon areas rich in metals (chrome, columbium, titanium) that are essential to the US defence industry, potentially curbing access to these important commodity frontiers. In this way, as Miller observes, 'climate-driven migration crises directly threaten powerful U.S. military-corporate business interests' (Miller 2017, p. 39). Not surprisingly, actors within the global extractive community have developed ways of managing this particular security issue. In *Militarized Global Apartheid* (2020) Catherine Besteman describes what she terms a 'new apartheid apparatus' created by the functioning core of a dispersed but interconnected 'global north' (p. 3). This apparatus is modelled on the pre-1990 South African system and 'takes the form of militarized border technologies and personnel, interdictions at sea, biometric tracking of the mobile, detention centres, holding facilities, and the criminalization of mobility' (p. 2). The primary function of this apparatus is, as it was for apartheid South Africa, to establish and maintain 'a labor regime responsive to the specific needs of industrial capitalism' (p. 14), at a time of 'advanced' (p. 1) neoliberal globalization.

---

[1] Several studies have pointed to the fact that the 1951 UN Convention needs to be updated to reflect the conditions currently affecting human and extra-human life on the planet. See Berchin et al. (2017).

I want to again stress the fact that while many of the migrants and refugees that are escaping eroded *social* worlds—and the poverty, crime, and formal or informal warfare that flourish in such social worlds—rather than depleted *ecological* worlds, *social and ecological exhaustion are consequences of the same extractive system.* As many of the critical interventions discussed in this book reveal, and as Moore's (2015) work on the material history of the biospheric crisis explains, this ecological crisis is not in itself the driver of social depletion and displacement, but rather a direct and inevitable consequence of the extractive class relations that have become ubiquitous through the expansion of the capitalist world-ecology. As stated in a different way by former Executive Secretary of the United Nations Framework Convention on Climate Change (UNFCCC) Christina Figueres (2022), the 'climate crisis, the nature crisis, the inequality crisis, the food crisis all share the same deep root: extractivism based on extrinsic principles' (p. xvii). This means that treating climate migration as a security problem is not simply a step towards demonizing migrants, it is a move that elides the entangled processes that cause displacement across the world-system.

However, and to return to the point made earlier in this chapter, from the perspective of the US Department of Defence and other institutions tasked with keeping the US secure, climate migration is an acute 'threat multiplier' (Marzec 2015, p. 2), both because of the stress migrants are seen to put on the system and because migrating people have been forced to abandon extraction sites crucial to (military and civilian) industry. Again, and as Ingrid Boas et al. observe in 'Climate Migration Myths', climate migration is described in securitization discourse as a 'looming security crisis' (2019, pp. 901–902). In this discourse, it is the climate migrant/refugee, rather than the extractive violence that has produced both biospheric breakdown and the precarious living conditions that are forcing people to become migrants/refugees, that is perceived as the agent of insecurity. When the US DoD is preparing to 'adapt' to climate change, this adaptation is dependent on this understanding of the climate migrant as a security problem. In this way, and as argued by Desmond Tutu, climate 'adaptation is becoming a euphemism for social injustice on a global scale' (2007, p. 166).[2]

---

[2] It should be noted that the American right wing has built a significant portion of its politics around this type of crisis and security imaginary. Donald Trump's promise to 'build the [already existing] wall' between the US and Mexico was introduced as an attempt

As the first chapters of this book touch upon, the securitization of migration in core or 'Western' societies has a long history intimately related to the formation and development of these societies. In the US, the construction of systems designed to encourage white Europeans to migrate to the US so that they could participate in settler colonialism or become (semi-skilled) labour in US-based industry, while at the same time limiting the possibility of people from other parts of the world from entering the nation (as in the Chinese Exclusion act of 1882), mark an important stage. Other notable developments include the notion that the post-Cold War era had inaugurated a clash of civilizations as proposed by Samuel Huntington (1996), the casting of Middle Eastern migrants as terrorists (Ceyhan and Tsoukala 2002), and the introduction of the idea that the US and the white European world are scenes where a 'great replacement' is taking place (Jones 2021). It is through such discursive, ideological, and judicial action that the migrant is rhetorically transformed from a human in need to a security concern.

THE MIGRANT AS MONSTROUS BODY IN THE AMERICAN CLIMATE EMERGENCY NARRATIVE

Many of these stages, and the extractive security logic that makes them seem rational, have been rehearsed in literature and other media forms. In early twentieth-century popular culture, the Mexican or Asian migrant is typically a fiendish other; a staggeringly cruel Spanish-speaking foil for the heroic white protagonist (Perry 2016), or a Fu Manchu-like criminal mastermind introducing illicit drugs and a culture of depravity into the white urban society of the West Coast (Mayer 2013). Since then, a long tradition of scholarship in cultural studies, postcolonial studies, American Studies, and related fields has problematized these representations.[3] While there are certainly still plenty of films and novels that rehearse this racist

---

to engineer securities vital to the effort to 'make America great again'. Throughout his presidency, but also before and after, the image of the migrant or refugee as a security threat—a criminal and a murderer, a terrorist, a burden on the (increasingly disman-tled) welfare state—circulated in American society (Chouhy and Madero-Hernandez 2019; Jones 2021). In this way, the climate migrant or refugee is transformed from a human in need into a concern to be managed by the departments and institutions tasked with national security.

[3] See, for instance, Pieterse (1992), Mueller et al. (2018), and Feagin (2020).

imagery in the open, the American Climate Emergency Narrative tends to avoid openly racist depictions of migrants and instead employ allegory through which black and brown bodies appear in the guise of undead or otherwise monstrous figures.

A striking example is Nicholas Sainsbury Smith's *Extinction Horizon* (2015).[4] In this novel, military scientists have crossed an experimental drug named VX-99 with the Ebola virus to create super soldiers of prodigious strength and endurance. Predictably, the virus gets loose and begins turning the human population on all continents into ferocious killers with superhuman strength and speed. These are referred to as 'Variants' and they are fast, hungry for human flesh and blood, and lack the capacity for complex thinking. In a matter of weeks, only 1 per cent of the human population remains worldwide. The US president succumbs to the infection and what is left of the American military retreats to ships and islands to begin the project of de-accelerating the planet-wide apocalypse and taking back the world. The protagonist of the novel is Master Sergeant Reed Beckham of 'Special Forces Delta Team Ghost'. He has spent much of his professional life killing people in the Middle East and to him, the Variants connote 'Indians' and the Middle Eastern 'insurgents'. In the novel and throughout the series, these racially inflected monsters flow into all the spaces capitalism has engineered: SUVs, aeroplanes, bungalows, shopping malls, skyscrapers, megacities, military facilities, and so on until capitalism teeters on the brink of extinction. If it were not for the joint effort of the military-industrial-medical complex, humanity would perish entirely under the unrelating waves of the radically mobile undead.

A similar story is told in director Marc Forster's Hollywood blockbuster *World War Z*.[5] The introductory credits of this film clearly signal that the film speaks about socio-ecological breakdown. This is done by

---

[4] *Extinction Horizon* is the first of an eight-book series called *Extinction: Season One*, penned by Sansbury Smith, first released in the form of self-published novels but now disseminated by Blackstone books in printed, electronic, and audiobook form. As Sansbury Smith explains in his author's profile, he used to be an employee of Iowa Homeland Security and Emergency Management, and he also claims that this background is important to his writing. The series is furthermore a part of a multi-authored 'cycle' also made up of the *Extinction: Dark Age, Extinction Survival, Extinction New Zealand,* and the *Redemption Trilogy*. The books of this cycle comprise, at the time of writing, twenty-two novels and two collections of short stories written by a number of different authors.

[5] This film is a reworking of Max Brooks sprawling, multi-perspective and faux-historical novel *World War Z* (2006) into a Hollywood action spectacular.

showing ominous images of various global cityscapes teeming with people that are then contrasted with sudden sequences that depict swarming ants, snarling, feeding alpha predators, flocks of birds, stranded dolphins and whales, and tempestuous oceans. In this way, the film calls attention to itself as a narrative about overpopulation, mobilities, 'natural' aggression, and (anthropogenic) climate violence.

The film moves from this collage of colliding human and animal worlds to protagonist Gerry Lane (Brad Pitt), who is making pancakes for his wife and two daughters in an idyllic suburban two-story house somewhere on the East Coast. Between Lane serving breakfast and doing the dishes, it becomes clear that this is not his natural habitat. He used to work for the UN in the most troubled parts of the Global South but has opted out so that he can tend to his growing, female-dominated nuclear family (Fig. 6.1).

After breakfast, the family heads into the city of Philadelphia, where they become stuck in traffic. This is part and parcel of urban life in the core, and the two daughters play happily in the back seat of their Volvo station wagon. Then, in a matter of seconds, a tide of bodies and teeth surges up further down the street, turning people into a rapidly growing horde of aggressive and border-crossing undead. Lane steals an abandoned RV and escapes the inner city, he loads up on water and medicine for his asthmatic daughter, and he is given shelter by a Latin American family in an apartment building in Newark. Lane manages to contact

Fig. 6.1 Gerry Lane and his family at a sunlit breakfast before the zombie migration crisis strikes in *World War Z*

friends at the UN and the US Department of Defense and because he is an unusually skilled operator, they send a helicopter for him and his family. While he waits, he tells the father of the host family that 'I used to work in dangerous places and people who moved survived, and those who didn't... Movement is life'. When the transportation arrives, Lane and his family escape by the skin of their teeth, undead bodies peeling off from the landing gears of the helicopter, but the family that has sheltered them remains behind. 'There is nowhere to go', the family's father exclaims.

While this father pays the ultimate price for his reticence, or inability, to move, Lane remains eminently mobile throughout the film. He travels to a crumbling South Korea and then on to Israel. Israel is the only nation in the film that has been able to control the infection. Endemically suspicious, they have been 'building walls there for two millennia', Lane is informed. Enormous ramparts reach up to the sky, encircling all of Jerusalem. Thus, movement is clearly also death, depending on which side of the wall you are standing. The walls at first keep the undead in check. However, 'mere islamophobic moments later' (Bould 2021, p. 30) a celebratory song by one of the Palestinian refugees catches the attention of the undead outside the walls. These undead look disturbingly like the stock image of the modern-day (climate) refugee from the Global South: emaciated, dressed in tatters, brown-skinned, and hungry. However, once they catch the scent or the sound of living humans, they become like the migratory, predatory animals or insects showcased during the film's introductory credits. Modelled by the special effects unit after 'animal-kingdom biomechanics', they scale the enormous walls with 'all the intuitive cooperation of an insect colony' (Boucher 2013). In other sequences, the undead move more like a tidal wave, flowing over and through everything in their path (Fig. 6.2).

So relentless is this wall-storming mass of brown bodies that all the guns of the Israeli armed forces cannot prevent them from breaching the barriers that keep the racialized infected apart from the armed and the vulnerable. Lane must keep on running. Movement is life.

The portrayal of the undead as seething and relentless animals or insects activates the planetary dimension discussed in detail in Chapter 4. Like the kaiju, these undead are vengeful avatars of the planet, and as such a security issue to resolve for US armed forces. In this way, the war that US troops engage in is a war with an angry planet; with a severely damaged and uncooperative ecology. At the same time, this imagery also informs the concurrent reading of the undead as *migrants* emerging from

Fig. 6.2 The undead climbing Israeli walls like teeming ants, and flowing over a bus like a tidal wave in *World War Z*

the peripheries of the world-system. As Mark Bould aptly argues in *The Anthropocene Unconscious* (2021), 'any depiction of a massive, mobile and "unwanted" population cannot not be about climate refugees' (p. 29). A similar reading of the undead in zombie films such as *World War Z* is performed by Penny Crofts and Anthea Vogl (2019), who observe that 'the anxieties connected to and generated by refugee movement reflect the monstrous qualities of the zombie' (p. 30). Thus, they argue, *World War Z's* depiction of a zombie war raises 'fears that refugee populations will contaminate systems of national order' (p. 30). With this in mind, it can be argued that what the US and Israeli soldiers are gunning down in scene after scene of *World War Z* is a nightmare evocation of a global precariat evicted by the extractive violence that capitalism has subjected their homes to.

By casting the migrant as a voracious zombie, the film accelerates the sense of acute crisis that is central to the American Climate Emergency Narrative. The undead is a figure that cannot be talked to, it scales any obstacle in its path to acquire entry into orderly, capitalist modernity, and, if given half the chance, will overflow and literally consume it. Simultaneously, Lane's small, nuclear family evoke capitalism as a soft, vulnerable, feminine, and white entity, something that must be made secure at all costs. The juxtaposition of this unit and the animalistic, biting undead legitimizes an immediate and total state of exception. Enormous walls and constant and extreme military-grade violence are now the order of the day. In this way, *World War Z* registers the precarity and violence that militarized capitalism produces in the US and globally only to narrate this violence as necessary and existential. The migrants as undead *must* be massacred as they scale the border. The unthinkable alternative is the grotesque and abject eating of Lane's wife and daughters.

Despite its celebration of brave and self-sacrificing soldier heroes, *World War Z* was not financially supported by the Department of Defense (Tarabay 2014). As an extremely expensive Hollywood production, it did, however, receive funding from other actors of the military-industrial and petrochemical complexes, this in exchange for various product placements. Thus, the film features bottles of Royal Purple Motor Oil, made by Calumet Specialty Products Partners (CLMT), a refiner and processor of hydrocarbon products headquartered in Indianapolis, Indiana. It also contains signs that advertise Capital One, an American bank headquartered in McLean, Virginia but operating throughout the US, as well as cans of Pepsi Cola that Lane, in a strange, drawn-out scene, guzzles down. However, the most striking example of product placement in the films is an automatic rifle that clearly reads ArmaLite. Inc, Geneseo, IL. US (Fig. 6.3).[6]

When named products are folded into the visual fabric of the story in this way, characters no longer drive cars, but a Mercedes, a Volvo, a Ford, or an Aston Martin. Similarly, soldiers do not simply fire automatic rifles or shotguns, but a Heckler & Koch, a Ruger, or, as in *Word War Z*, an ArmaLite. Such visualizations can be understood as an effect of the prevalent integration of the entertainment industry, the (military)

---

[6] Numerous recognizable guns are shown in the movie. For a complete list, see https://www.imfdb.org/wiki/World_War_Z_(2013). Apart from ArmaLite, it is difficult to say which manufacturers that actually paid to showcase their weapons.

Fig. 6.3   ArmaLite product placement in *World War Z*

industrial complex, and the spoken and unspoken geopolitical ambitions of the core. When Lane hunkers down among clearly branded containers of motor oil, or when soldiers flaunt branded assault rifles, the film does not simply register the fact that the world Lane exists within was made through oil and guns, it actively promotes a specific actor/corporation that is trying to ensure and expand its life span within this world. In other words, the hope is that a cinematic gunning down of the undead in the film should stimulate the sales of hydrocarbons, sugars dissolved in water, mortgages, and assault rifles from specific agents, this while projecting a general sense that these items and services will remain essential even as the world burns. In this way, narratives from the core do not just celebrate the sense that the core is essential to the prolongation of human life, they advertise and seek to stimulate specific actors and corporations. This is part of how capitalism makes fiction within the world-system, and also of how fiction sponsored by actors within the system attempts to make the world.

In *World War Z*, this network of commercial entities helps make possible, even accelerates, the film's anti-migration narrative. Corporations such as these rely upon, feed from, and help maintain the very intellectual and physical borders that also secure US world-system hegemony. Petrochemical industries, banks, and arms manufacturers can tolerate, even be energized by, capitalist crisis,[7] but they prefer labour to remain in place and is premised, in ways described by Cedric Robinson (2019), upon a racialization of labour. It is not strange, then, that much popular culture from the core, energized by product placement, casts displaced and supremely mobile racialized labour as deeply disturbing, or that the anxieties such disturbance produces are promoted via images of gruesome violence. Again, as Besteman (2020) and Miller (2017) note, for commodities such as palm oil, sugar, or lithium to flow into the industries of the core, racialized labour must be kept in their place. At the same time, fear that such labour might be breaching borders is what energizes xenophobia, prepper lifestyles, and gun sales. *World War Z* thus registers, but also contributes to, this precise confluence between the needs and commodifiable fears of the core of the capitalist world-system.

THE BODY IN PLACE

If movement is life, as Lane argues in *World War Z*, it is clearly also a privilege. When the father of the Latin American family that harbours Lane's family for the night exclaims that 'There is nowhere to go', he speaks from a position of peripherality. For him, this is the end of the road. This does not mean that there is nowhere to go, as Lane's family's last-minute helicopter transfer to the aircraft carrier demonstrates, but such mobilities are reserved for those belonging to the predominantly white core of the world-system; those who are defined as fully human.[8] Lane relies on such mobilities and, throughout *World War Z*, he never stops his movement. He is shepherded by soldiers and civilians all over the world. He takes control of vehicles, he is picked up by helicopters, Air Force transportation planes, and jet airliners, in the process crossing all conceivable national and intrastate borders. In these fast-moving, border-transcending vehicles, Lane is a *body in place*. By contrast, the racialized migrant, whether cast as covering Latin American or ravenous undead, is

---

[7] As noted by Jennifer Carlson (2015), gun sales tend to go up at times of perceived economic and political decline.

[8] An important point made my Moore (2016) is that extractive capitalism has thrived partially through excluding a large portion of humanity from the category of the human so that 'indigenous peoples, enslaved Africans, nearly all women, and even many white-skinned men (Slavs, Jews, the Irish)' (p. 79) could be regarded as Nature rather than as people. Treated as such, they had no claim to human rights and their labour could be easily and violently extracted.

what Sarah Ahmed has called a 'body out of place' (2013).[9] In concert with Judith Nicholson (2016), Paul Gilroy (2020), and Barbara Harris Combs (2022), Ahmed notes how anti-black racism identifies the body of colour as out of place at the moment when it tries to cross into, or when it presumes to exist within, spaces colonized by whiteness. This is what motivates the immense violence enacted in the film. The racialized migrant turned undead is monstrously out of place in the white spaces produced by capitalist modernity. Thus, there is no alternative to the brutal violence used to keep the undead at bay in *World War Z*.

As discussed in previous chapters of the book, the notion that some bodies are out of place is ingrained in the extractive history of the US's rise to world-system hegemony. Chapter 2 thus shows how settler writing narrated the mobility of the invasive white body as natural, irresistible, and transcendent, while Indigenous mobility was depicted as disturbing and dangerous. Similarly, as discussed in Chapter 3, the movement of black bodies was strongly regimented during slavery, and the fugitive slave act allowed slave owners to pursue black bodies far into 'free' territory. In the apartheid era that followed the 14th amendment, the social relationship between black and white bodies was determined through the legal and semi-legal designation of particular spaces as white or black. The black codes that emerged in the South immediately after the Civil War thus sought to hinder black people from leaving the plantation, from congregating, and from entering white spaces. After the Supreme Court gave its blessing to segregationist policies in their decision on the Homer Plessy v. Ferguson case in 1896, restaurants, rest rooms, parks, schools, compartments on public transportation, health-care facilities, and so on could be, and frequently were, marked as white or black.

The Brown v. Board of Education of Topeka case of 1954 helped make this overtly racist practice illegal, but the marking of certain bodies as out of place (and others as in place) remains systemic in American (and much of global) society. As discussed by Simone Browne (2015), present-day 'racialized disciplinary society' (p. 6) locates white, brown, and black bodies differently in relation to space, it surveilles the movement of bodies, and it curbs or privileges this movement according to race. The very well-documented profiling of black and brown people driving cars or attempting to enter the US are present-day examples of how this society

[9] As Judith Nicholson and Paul Gilroy have observed, Black mobilities remain severely constrained in the US and other parts of the world.

operates (Jernigan 2000; Gilroy 2020; Romero 2008; Ramirez et al. 2003). This situation, it should be added, is an important part of how what Robinson (2019) terms racial capitalism, and Besteman (2020) calls militarized global apartheid, works. Just like the US cotton plantation demanded the presence of a vast (racialised) workforce, the extraction and refinement of resources in Sub-Saharan Africa require a stationary labour force: a black or brown body *in place*. In this way, 'racialized disciplinary society' is central to the working of the capitalist world-ecology.

It is thus not surprising that the colour coding of (migrant) bodies informs the American Climate Emergency Narrative's account of the lived experience of biospheric erosion and its impact on human and extra-human worlds. As Ahmed, Browne, and other scholars argue, the dominant and racialized 'disciplinary society' relies on a biopolitics that surveilles and calls attention to the body as a racialized (and gendered) entity, and that encourages, sanctions, or enforces certain actions upon this body if it is perceived to transgress. In texts that take place in futures where the biosphere is breaking down and people are forced to take to the road, the disciplining of bodies provides the narrative with a certain structure. To return to *World War Z*, the racialized undead become legitimate targets of violence. The relative blackness of the undead body marks it as an entity that must be prevented from crossing the border that separates the periphery from the core. This is part of the extractive logic that provides the film with its narrative momentum. It can be argued that Lane's assignment is ultimately to ensure that bodies such as his remain mobile and in place. By contrast, Lane's (normative settler) whiteness provides him with a universal passport. His body is in place everywhere. Yet, such privileged white mobilities demand the fortification of borders that curb the mobilities of those racialized, gendered, sexualized, or classed people who are described as not entirely human.

In *World War Z*, Lane manages to halt the unfolding apocalypse. His trick is to infect himself with a conventional virus that makes him a bad host for the zombie virus. Thus contaminated, he is of no interest to the undead. Lane can hide in their midst, seemingly undead but in fact as mobile as ever. His discovery creates breathing space for survivors and the military all over the world. In the final sequence of the film, images of vaccines being dropped from military transport planes, evacuees walking unmolested past zombies, soldiers showering hordes of the undead with napalm, and bomb planes targeting mass gatherings of zombies are interspersed with scenes that show how Lane arrives on a boat to be reunited with his family. His daughters and his wife hug him and celebrate his return with tears and shouts of joy. They are grateful he has returned in one piece, but also because he has restored their future. Even though their 'war has just begun' as Lane thinks to himself, the ending promises that it will be possible to restore the world that provided such tranquil privilege and such effortless mobilities in the opening kitchen scene.

THE UNBEARABLE WHITENESS OF THE AMERICAN CLIMATE EMERGENCY NARRATIVE

World War Z imagines a future where white agents such as Lane are able to assist the military-industrial-science complex in preventing the (gothicised) climate emergency from unfolding. As a result, the white, middle-class body out of which subjectivity flows will be able to reclaim its privileges and retain its transcendent mobilities. But this is not the only possible trajectory. Other American Climate Emergency Narratives are less optimistic and explore futures where the gallant efforts of men like Lane have failed and where, suddenly, the white body finds itself *out of place* and deprived of all the comforts that it once enjoyed. The main characters of these texts are typically characters who used to have a place at the core but who find themselves in the shoes of the undead in *World War Z*: they cannot get on planes or drive cars, they are constantly hungry and miserable, and they may even suffer a kind of systemic violence. Cormac McCarthy's influential, post-apocalyptic *The Road* (2006), discussed in more detail in Chapter 7, makes this exact point when, in a retrospective scene, the wife of the novel's (white) male protagonist informs him that 'We're not survivors. We're the walking dead in a horror film' (p. 47). In *The Road*, the whiteness that clings to the wandering father and his son means nothing and thus offers no protection.

By revoking the privileges of white people, narratives such as *The Road* imagine futures where the social relationships that Lane works so hard to maintain in *World War Z* have been permanently terminated. This does not mean, however, that *The Road* is antithetical to *World War Z*. Rather, McCarthy's novel takes place in a future world in which agents such as Lane has failed to prevent the climate apocalypse from unravelling so that the utter chaos that threatens on the other side of the nightmare created by the film has become a reality. In *World War Z*, the ultimate horror is not the emergence of a global and enormously violent pandemic, but the possibility that Lane's and his family's white, middle-class bodies should cease to generate privilege. Such cessation would mean the collapse of the capitalist world-system they are part of and that has privileged them. Again, in *The Road*, this is precisely what has happened. The white bodies of the father and the son mean nothing and thus offer no protection. Thus, both texts can be said to explore the same apocalyptic future but to focus on two different stages. *The Road* obviously eschews the spectacular fireworks and the hopeful closure of *World War Z*, yet it makes use of a similar climate emergency imaginary where the collapse of capitalist modernity inaugurates a socio-ecological state so bleak it cancels life entirely.

There are a number of studies of 'climate fiction' that focus on, and recommend, precisely this rhetorical scheme. The argument put forward by these studies is that 'climate fiction' such as *The Road* raises awareness of global warming by allowing readers to imaginatively and emotionally enter the dark social and ecological futures that the 'climate crisis' will produce. Thus, as discussed in the introduction, Gregers Andersen attempts, in his book *Climate Fiction and Cultural Analysis* (2020),

> to demonstrate that climate fiction represents a vital supplement to the reports published by the United Nations' Intergovernmental Panel on Climate Change (IPCC) because, by depicting humans in worlds resembling those forecast by the IPCC, climate fiction provides speculative insights into how it might be to feel and understand in such worlds. And these are basically insights we as contemporary humans cannot obtain anywhere else. (p. 1)

Also following this line of thinking, Alexa Weik Von Mossner (2017) has proposed that 'climate fiction' such as Paolo Bacigalupi's *The Water Knife* 'channels its dramatic evocation of climate change's external effects through the minds and bodies of its protagonists, thus allowing readers to imaginatively experience *what it is like* to live in a climate-changed world' (p. 174, italics in the original).[10] In other words, fiction such as *The Road* or *The Water Knife* narrate how a certain, previously privileged social stratum (the 'contemporary humans' that Andersen describes as reading 'climate fiction') wakes up to a world where nothing or too little has been

---

[10] Similarly, Paolo Bacigalupi has argued that he wrote the novel *The Water Knife* as a response to America's 'willingness to pretend that climate change wasn't real, and wasn't a pressing problem for us' (Bacigalupi 2017).

done to prevent global warming and where their privileges and mobilities have been drastically reduced as a consequence. By telling such stories, this fiction makes people who belong to this still privileged stratum aware of the profound social upheaval that may result if the ecology of the planet is allowed to continue eroding. What 'climate fiction' furnishes is not simply an intellectual wake-up call, but an emotional and affective encounter with this deeply disturbing future.

However, while this type of affective engagement may well serve a purpose, stories such as the ones promoted by Andersen or Mossner risk eliding the temporal, social, and geographical dispersion of global warming, other types of biospheric erosion such as pollution and the planetary spreading of plastic waste, and, most importantly, the social and economic drivers that have caused these developments. To return to the beginning of this chapter, and also to the historical perspective that informs this book, biospheric breakdown is, to a large number of people, not a thing that will unfold in the future, but an *ongoing socio-ecological disaster*. Again, as Rob Nixon shows in his ground-breaking book *Slow Violence and the Environmentalism of the Poor* (2011) and as Farhana Sultana (2022) has also observed, climate breakdown is in fact a *socio-ecological* development and thus an already lived experience in many parts of the world. The existing capitalist world-ecology was built upon a foundation of theft of Indigenous lands and on the violent extraction of nature and people. This enabled a thriving modernity at the core and in parts of the global semiperiphery, but it also produced profound inequality, soil erosion, pollution, war, global warming, and, of course, displacement. When what Andersen and Mossner refer to as 'climate fiction' focuses on the loss of privileges in the Global North, rather than the history of extraction, the slow violence and the long-established inequalities that exist in the Global South and that have made these privileges possible, a crucial story goes untold.

The absence of this story, and the history of extraction it would tell, in fiction that seeks to make 'contemporary humans' aware of global warming by focusing on the curbing of (white) mobilities and privileges is noticeable. One of many examples is Brian Hart's *Trouble No Man* (2019). This text narrates, in a series of temporally fragmented chapters, the life of the white Roy Bingham, once a semi-professional skateboarder, who tries to settle down in rural California with his partner Karen and their two daughters. This is difficult not only because Roy is an itinerant soul—he craves movement on his skateboard, in his car, or motorcycle down the road and across state and nation-state borders, and he shuns commitment to people and places—but because increasing drought is eroding both the land and the politics of his state. New and violent communities made up of white farmers who have lived on the land 'forever' (p. 105) begin to take command of the area. These farmers are running out of water, and their efforts to seize power are made easier by the fact that they have a tradition of sending their sons and daughters to the military to become trained militia. These new 'totally racist, not to mention fascist' (p. 105) communities attract preppers from all over the US, and they have begun to form militant enclaves that unite under the 'Jeffersonian flag' or similar signs of secession. They erect fences and establish political borders and roadblocks and while people of colour are even more likely to be stopped at these improvised borders, Roy, Karen, and their two white daughters and friends also find their mobilities constrained. Going for a drive one evening, they come across newly formed borders, and there is no guarantee that they will be allowed to pass: 'The Jeffs have doubled their security at the roadblocks, and after the first two they decide to turn around and go home. More and more they're in occupied territory. Strangers in camo with M4s directing traffic' (pp. 129–130). In this future, the white bodies of the Binghams do not guarantee entry any longer. The new and increasingly politicized security order has transformed him and his family into *bodies out of place*. Thus, and as in *The Road*, this imagined future is hostile not simply because the land is unable to yield the clean water and the food needed to sustain human bodies, but primarily because the borders that regulate human mobilities have become difficult to cross for the white subject.

Tensions rise further and the part of California where the Binghams live is increasingly riddled with fences, roadblocks, and landmines. Still mobile, the grown daughters manage to escape to Alaska. In futures transformed by global warming, northern territories are often described as offering sanctuary. Karen and Roy find moving more difficult, partly because of the borders put in place, but also because it is hard and expensive to abandon land that you formally own. As they make plans, Karen and Roy are persuaded by Karen's old stepfather to transgress one of the new borders, to enter space that has become forbidden to them. They are attacked, and manage to escape, but are now marked as enemies of the local clan. Before long, Karen has been shot and killed during an ambush. In this future transformed by socio-ecological breakdown, it is not just South American migrants or the undead that are kept in their place, and who pay with their lives if they go astray. Roy responds to this killing via the same logic. The entire novel is thus framed by a scene where Roy and another man take vengeance for the killing of Karen. Following the established paradigm, biospheric erosion produces both emergency and the state of exception that suspends the rule of law. The difference is that, in *Trouble No Man*, white people are not only the engineers but also the victims of such improvised states of exception. All mobilities are curbed and everyone can be the target of organized gun violence.

Unlike many other American Climate Emergency Narratives, *Trouble No Man* is not entirely oblivious to systemic violence. Watching California drying out and filling up with armed and border-building secessionists, Karen reflects that 'This has all happened before, right? [...] Drought, fire, strife, dead forests, the threat of world war. Actually, this has happened forever. It's never stopped. Moves continent to continent. It'll be our turn soon but not yet' (267). The observation that biospheric erosion and war go hand in hand and that they are global phenomena does suggest an understanding of socio-ecological breakdown as world-systemic, but the claim that such breakdown 'has happened forever' erases the particular history of violence that the emergence of the world-system initiated. Thus, Karen's statement rehearses the Anthropocene thesis: the human species has always made the world burn. Perceived via the logic this thesis affords, capitalist modernity is not the engine of 'fire, strife, dead forests, the threat of world war', it is rather what has, until now, protected people like Karen, Roy, and the preppers that surround them, from being engulfed. It is the accelerating erosion of this modernity that makes it 'our turn', as Karen puts it. The roadblocks and border skirmishes that eventually kill Karen are only the beginning—but it is the beginning of a development located, for white Californians, in the future.

Trouble No Man obviously lacks the spectacular and frequently racist cinematics of *World War Z*, and Roy is a far less heroic character than Brad Pitt's Lane or even the father of McCarthy's *The Road*. Yet, this novel also, if more furtively, centres the white body as a prioritized victim of socio-ecological breakdown. This is the entity imagined as experiencing the detrimental ecological, social, and material effects that biospheric breakdown causes. The book also locates such breakdown to the future. In this way, *Trouble No Man* also contributes to the erasure of the ongoing, slow, and fast violence that extractive capitalism has long subjected people in the periphery and semiperiphery to.

Observing these Eurocentric/Amerocentric tendencies in American 'climate fiction', Matthew Schneider-Mayerson (2019) has noted a lack of, or disinterest in, 'climate justice' and how the prevalent description of 'climatic destabilization primarily as a problem for white, wealthy, educated Americans' (p. 945) normalizes a white experience of the world while at the same time universalizing the white subject into a representative of the suffering human. Again, this representation erases the long history of climate injustice as lived by people of colour, in the process potentially reifying 'a narcissistic tendency among many white American readers' (p. 945). Similarly, Hsuan L. Hsu and Bryan Yazell (2019) have argued that fictions about future biospheric erosion that centre the suffering of white middle-class protagonists, while ignoring the long history of extractive colonialism that has produced the ecological crisis that drives the narrative, exemplify what they term 'structural appropriation' (p. 347). This is 'a process in which the world-threatening structural violence that has already been experienced by colonized and postcolonial populations is projected onto American (and predominantly white) characters and readers' (p. 347). *Trouble No Man* exemplifies this tendency.

Thus, while *Trouble No Man* does not locate the emergency in the movement of black and brown bodies across white borders, it does rehearse the notion that socio-ecological breakdown is a crisis for a particular white subject. In this way, its description of the lived experience of future climate breakdown has more to do with the anxieties that haunt the core, than with the processes that have caused breakdown in the stories, or with the unevenness of its impact. Again, little is said in this novel (or in *The Road*) about why the land is drying out and dying, and why the services of the world-system have been partially or completely suspended. Instead, socio-ecological breakdown is registered via certain *effects* that breakdown will have on the white American subject. Such subjects, it is suggested, risk being forced to leave the land they once settled, and they may have to undertake fundamentally unsafe, migratory journeys towards uncertain or even non-existent goals. In *Trouble No Man*, this uncertainty is at the heart of the emergency the novel describes.

DISPLACEMENT AND INSECURITY

This chapter begins and ends with the realization that the ongoing fortification of the physical and legal barriers that keep climate migrants/refugees displaced is a manifestation of both militarized global apartheid as defined by Besteman (2020) and of environmentality as described by Marzec (2015). Rather than trying to come to terms with the deep and underlying systemic processes that cause global warming, states in the Global North are treating the socio-ecological upheaval as a national security issue that can only be resolved with the help of the existing national security apparatus. In the process, people displaced by the planetary emergency become a 'rationale for measures to strengthen and protect national and regional borders in the Global North' (Boas et al. 2019, p. 902). Ultimately, border building and border enforcing in the Global North is an effort designed to keep extractive, racial capitalism operational, in the hope that capitalism will somehow become so sustainable that no actual systemic change will be needed to avert socio-ecological collapse, or simply to postpone the collapse for a few more years.

The American Climate Emergency Narrative registers this particular development through two dominant tropes. As the analysis of *World War Z* reveals, a strongly speculative register is employed to cast the climate migrant or refugee as a security threat of such bizarre proportions that it can only be kept at bay through large-scale military action. Thus, the film leverages military violence as a rational way to manage the displacement of large groups of people produced by ongoing planetary breakdown. In this way, displacement is narrated as yet another security crisis akin to that presented by a riotous planet in Chapter 4, or by competing states within the world-system as in Chapter 5. The implicit message of this narrative, and many others like it, is that biospheric erosion will produce conditions where capitalist modernity finds itself under siege. At the same time, climate deterioration will turn neighbours into displaced and riotous monsters. This again produces socio-ecological breakdown as an emergency for capitalist modernity and thus legitimizes the state of exception that allows for constant and ferocious military violence.

Texts such as *The Road* and *Trouble No Man* complicate this very simplistic casting of the climate migrant as a transgressive monster. In these texts, the climate migrant is instead a representative of white lower-middle-class society. These narratives change the story by centring on the people who have been displaced by socio-ecological breakdown, rather than those who seek to prevent mobility. By casting the migrant as a recognizable person from the present core or US semiperiphery, readers are asked to put themselves into the worn shoes of the displaced. This is what your life may be like if nothing is done about the climate emergency, these texts are saying. The general assumption, rehearsed by much climate fiction scholarship (Von Mossner 2017; Andersen 2020), is that such warnings may inspire change on some level. The reader will want to avert the catastrophe narrated on the pages of these dystopian novels. There can be little doubt that the reading of *The Road* and *Trouble No Man* produces an awareness of climate breakdown as a phenomenon. These texts may even encourage some kind of climate activism. However, as I have argued, while these texts do transport their audiences and readers into violent futures where climate breakdown has caused the partial collapse of the nation-state and where (white) people have lost their settler privileges, they do not recognize how the suffering experienced by their protagonists has long been lived by other global communities, and they do not explain the extractive history that has produced this present and future suffering in any detail. Race-based, myopic politics may loom in the background as an engine of violence in stories such as *Trouble No Man*, but none of the characters in the novel work towards upsetting the system that has produced climate breakdown. Via the micro-perspective employed in the novel, the characters are confined to ending their displacement by escaping the zone of suffering. This is a short-term strategy that dismisses any other relationship to the environment than the one made possible by capitalism. This is how the individuals in these texts adjust to climate breakdown: not by changing things but by *adapting* to new circumstances; by moving on, gun in hand, until the gates of capitalist society are again opened and the individual is let in, for now. As fiction from the core, these are the futures these texts are conditioned to conceive. In this way, the narratives discussed in this chapter all imagine worlds where US capitalism is still in place. It may be predatory and unjust, but it still exerts an enormous gravity on the displaced characters of the texts. Finding themselves outside of its cold embrace, and stripped of its convenient means of transportation, there is nowhere else to go.

WORKS CITED

Ahmed, Sara. 2013. *Strange Encounters: Embodied Others in Post-Coloniality*. London: Routledge.

Andersen, Gregers. 2020. *Climate Fiction and Cultural Analysis: A New Perspective on Life in the Anthropocene*. London: Routledge.

Apap, Joanna, and Capucine du Perron de Revel. 2021. *The Concept of 'Climate Refugee': Towards a Possible Definition*. European Parliament.

Bacigalupi, Paolo. 2015. *The Water Knife*. New York: Knopf Doubleday

———. 2017. 'I Wrote the Water Knife.' Facebook Post, January 31. https://www.facebook.com/PaoloBAuthor/posts/i-wrote-the-water-knife-because-i-was-concerned-about-americas-willingness-to-pr/10154946832703904/.

Berchin, Issa Ibrahim, Isabela Blasi Valduga, Jéssica Garcia, and José Baltazar Salgueirinho Osório de Andrade. 2017. 'Climate Change and Forced Migrations: An Effort Towards Recognizing Climate Refugees.' *Geoforum* 84: 147–150.

Besteman, Catherine. 2020. *Militarized Global Apartheid*. Durham, NC: Duke University Press.

Boas, Ingrid, Carol Farbotko, Helen Adams, Harald Sterly, Simon Bush, Kees Van der Geest, Hanne Wiegel, Hasan Ashraf, Andrew Baldwin, and Giovanni Bettini. 2019. 'Climate Migration Myths.' *Nature Climate Change* 9 (12): 901–903.

Boucher, Geoff. 2013. '*World War Z* Cover Story.' *Entertainment Weekly*, March 29. https://ew.com/article/2013/03/29/world-war-z-cover-story/.

Bould, M. 2021. *The Anthropocene Unconscious: Climate Catastrophe Culture*. London: Verso.

Brooks, Max. 2006. *World War Z: An Oral History of the Zombie War*. New York: Crown.

Browne, Simone. 2015. *Dark Matters: On the Surveillance of Blackness*. Durham, NC: Duke University Press.

Carlson, Jennifer. 2015. *Citizen-Protectors: The Everyday Politics of Guns in an Age of Decline*. Oxford: Oxford University Press.

Ceyhan, Ayse, and Anastassia Tsoukala. 2002. 'The Securitization of Migration in Western Societies: Ambivalent Discourses and Policies.' *Alternatives* 27: 21–39.

Chouhy, Cecilia, and Arelys Madero-Hernandez. 2019. '"Murderers, Rapists, and Bad Hombres": Deconstructing the Immigration-Crime Myths.' *Victims & Offenders* 14 (8): 1010–1039.

Combs, Barbara Harris. 2022. *Bodies Out of Place: Theorizing Anti-Blackness in US Society*. Athens, GA: University of Georgia Press.

Crofts, Penny, and Anthea Vogl. 2019. 'Dehumanized and Demonized Refugees, Zombies and World War Z.' *Law and Humanities* 13 (1): 29–51.

d'Appollonia, Ariane Chebel. 2017. *Frontiers of Fear: Immigration and Insecurity in the United States and Europe*. Ithaca: Cornell University Press.

De Bruyn, Ben. 2020. 'The Great Displacement: Reading Migration Fiction at the End of the World.' *Humanities* 9 (1): 25.

Feagin, Joe R. 2020. *The White Racial Frame: Centuries of Racial Framing and Counter-Framing*. London: Routledge.

Figueres, Christina. 2022. 'Foreword.' In *Earth for All: A Survival Guide for Humanity*, edited by Sandrine Dixson-Declève, Owen Gaffney, Jayati

Ghosh, Jørgen Randers, Johan Rockström, and Per Espen Stocknes, xvii–xviii. Gabriola Island: New Society Publishers.

Foster, Marc, director. 2013. *World War Z*. Paramount Pictures.

Gilroy, Paul. 2020. 'Driving While Black.' In *Car Cultures*, edited by Daniel Miller, 81–104. London: Routledge.

Haraway, Donna J. 2016. *Staying with the Trouble: Making Kin in the Chthulucene*. Durham: Duke University Press.

Hart, Brian. 2019. *Trouble No Man*. New York: HarperCollins.

Hsu, Hsuan L., and Bryan Yazell. 2019. 'Post-Apocalyptic Geographies and Structural Appropriation.' In *Routledge Companion to Transnational American Studies*, edited by Nina Morgan, Alfred Hornung, Takayuki Tatsumi, 347–356. Abingdon: Routledge.

Huntington, Samuel P. 1996. *The Clash of Civilizations and the Remaking of World Order*. New York: Simon & Schuster.

IEP. 2020. *Over One Billion People at Threat of Being Displaced by 2050 Due to Environmental Change, Conflict and Civil Unrest*. https://www.economicsandpeace.org/wp-content/uploads/2020/09/Ecological-Threat-Register-Press-Release-27.08-FINAL.pdf.

Jernigan, Adero S. 2000. 'Driving While Black: Racial Profiling in America.' *Law & Psychology Review* 24: 127–138.

Jones, Reece. 2021. *White Borders: The History of Race and Immigration in the United States from Chinese Exclusion to the Border Wall*. Boston, MA: Beacon Press.

Marzec, Robert P. 2015. *Militarizing the Environment: Climate Change and the Security State*. Minneapolis: University of Minnesota Press.

Mayer, Ruth. 2013. *Serial Fu Manchu: The Chinese Supervillain and the Spread of Yellow Peril Ideology*. Philadelphia: Temple University Press.

McCarthy, Cormac. 2006. *The Road*. New York: Alfred A Knopf.

Miller, Todd. 2017. *Storming the Wall: Climate Change, Migration, and Homeland Security*. San Francisco, CA: City Lights Books

Moore, Jason W. 2015. *Capitalism in the Web of Life: Ecology and the Accumulation of Capital*. London: Verso.

———. 2016. 'The Rise of Cheap Nature.' *Anthropocene or Capitalocene? Nature, History, and the Crisis of Capitalism*, edited by Jason W. Moore, 78–115. Oakland: PM Press.

Mueller, Jennifer C, Apryl Williams, and Danielle Dirks. 2018. 'Racism and Popular Culture: Representation, Resistance, and White Racial Fantasies.' In *Handbook of the Sociology of Racial and Ethnic Relations*, edited by Pinar Batur and Joe R. Feagin, 69–89. Chamalthusser: Springer.

Nicholson, Judith A. 2016. 'Don't Shoot! Black Mobilities in American Gunscapes.' *Mobilities* 11 (4): 553–563.

Nixon, Rob. 2011. *Slow Violence and the Environmentalism of the Poor*. Cambridge, MA: Harvard University Press.

Padrón, Ryan S., Lukas Gudmundsson, Bertrand Decharme, Agnès Ducharne, David M. Lawrence, Jiafu Mao, Daniele Peano, Gerhard Krinner, Hyungjun Kim, and Sonia I. Seneviratne. 2020. 'Observed Changes in Dry-Season Water Availability Attributed to Human-Induced Climate Change.' *Nature Geoscience* 13 (7): 477–481.

Perry, Leah. 2016. *The Cultural Politics of U.S. Immigration: Gender, Race, and Media*. Vol. 17. New York: New York University Press.

Pieterse, Jan Nederveen. 1992. *White on Black: Images of Africa and Blacks in Western Popular Culture*. New Haven: Yale University Press.

Ramirez, Deborah A., Jennifer Hoopes, and Tara Lai Quinlan. 2003. 'Defining Racial Profiling in a Post-September 11 World.' *American Criminal Law Review* 40 (3): 1195–1233.

Robinson, Cedric J. 2019. *On Racial Capitalism, Black Internationalism, and Cultures of Resistance*. London: Pluto Press.

Romero, Mary. 2008. 'Crossing the Immigration and Race Border: A Critical Race Theory Approach to Immigration Studies.' *Contemporary Justice Review* 11 (1): 23–37.

Sansbury Smith, Nicholas 2015. *Extinction Horizon*. New York: Orbit.

Schneider-Mayerson, Matthew. 2019. 'Whose Odds? The Absence of Climate Justice in American Climate Fiction Novels.' *ISLE: Interdisciplinary Studies in Literature and Environment* 26 (4): 944–967.

Seneviratne, Sonia, Neville Nicholls, David Easterling, Clare Goodess, Shin-jiro Kanae, James Kossin, Yali Luo, Jose Marengo, Kathleen McInnes, and Mohammad Rahimi. 2012. 'Changes in Climate Extremes and their Impacts on the Natural Physical Environment.' In *Managing the Risks of Extreme Events and Disasters to Advance Climate Change Adaptation*, edited by C. B. Field, V. Barros, T. F. Stocker, D. Qin, D. J. Dokken, K. L. Ebi, M. D. Mastrandrea, K. J. Mach, G.-K. Plattner, S. K. Allen, M. Tignor, and P. M. Midgley. A Special Report of Working Groups I and II of the Intergovernmental Panel on Climate Change (IPCC), 109–230. Cambridge: Cambridge UP.

Sultana, Farhana. 2022. 'The Unbearable Heaviness of Climate Coloniality.' *Political Geography* 99: 102638.

Tarabay, Jamie. 2014. 'Hollywood and the Pentagon: A Relationship of Mutual Exploitation.' *Aljazeera America*, July 29. http://america.aljazeera.com/art icles/2014/7/29/hollywood-and-thepentagonarelationshipofmutualexploit ation.html.

Tutu, Desmond. 2007. 'We Do Not Need Climate Change Apartheid in Adaptation.' *Human Development Report 2007*, 166–186. New York: United Nations.

Vietti, Francesca, and Todd Scribner. 2013. 'Human Insecurity: Understanding International Migration from a Human Security Perspective.' *Journal on Migration and Human Security* 1 (1): 17–31.

Von Mossner, Alexa Weik. 2017. 'Sensing the Heat: Weather, Water, and Vulnerabilities in Paolo Bacigalupi's *The Water Knife*.' In *Real-Yearbook of Research in English and American Literature: Vol. 33 (2017): Meteorologies of Modernity. Weather and Climate Discourses in the Anthropocene*, edited by Sarah Fekadu, Hanna Straß-Senol, and Tobias Döring, Vol. 33, 173–190. Tübingen: Narr Francke Attempto Verlag.

Ummenhofer, Caroline C, and Gerald A Meehl. 2017. 'Extreme Weather and Climate Events with Ecological Relevance: A Review.' *Philosophical Transactions of the Royal Society B: Biological Sciences* 372 (1723): 20160135.

Wyett, Kelly. 2014. 'Escaping a Rising Tide: Sea Level Rise and Migration in Kiribati.' *Asia & the Pacific Policy Studies* 1 (1): 171–185.

Xu, Chi, Timothy A. Kohler, Timothy M. Lenton, Jens-Christian Svenning, and Marten Scheffer. 2020. 'Future of the Human Climate Niche.' *Proceedings of the National Academy of Sciences* 117 (21): 11350–11355.

Yazell, Bryan. 2020. 'A Sociology of Failure: Migration and Narrative Method in US Climate Fiction.' *Configurations* 28 (2): 155–180.

CHAPTER 7

Ruins

THE END OF THE WORLD

All texts discussed in this book express concerns about the damage done to the planet, and some even display a certain awareness of the fact that extractive, militarized capitalism has produced this damage. Even so, as I have argued, the plot that drives the American Climate Emergency Narrative is typically geared towards preserving or restoring capitalism and the (racialized and gendered) society that it has produced and operates through. The monocultural or capitalist realist storyworlds in which these texts play out typically make any other outcome except restoration or preservation literally unthinkable. Even in the few texts that envision futures where the US has been ousted as global hegemon, capitalism and the capitalist world-system are still never in doubt. In this way, narratives from the core typically insist that militarized capitalist modernity will be able to (violently) resolve the various ecological and human challenges that it has helped produce, or at least provide some kind of shelter for those entitled to it, even if the US loses its privileged control of the world-system.

This chapter turns to a final type of American Climate Emergency Narrative that is less optimistic in its view of capitalism's prospects, that envisions futures where capitalism is failing, or that takes the reader into the seemingly unimaginable territory where capitalist modernity is truly over. Like most of the fiction studied so far, these texts centre on white

© The Author(s) 2024
J. Höglund, *The American Climate Emergency Narrative*,
New Comparisons in World Literature,
https://doi.org/10.1007/978-3-031-60645-8_7

and formerly middle-class protagonists, but the stories they tell play out in futures where capitalism has not survived the ecological and economic crisis it has caused. In these futures, the economic, military, and social infrastructure capable of extracting Cheap Nature has collapsed entirely. There are no commodity frontiers left that can re-energize the world-system and the epochal crisis that capitalism and the planet are facing in the present moment has run its course. The narratives that tell these stories are not just post-apocalyptic, but formally post-capitalist in the sense that the only things that remain are the ruins of capitalist society: architecture in the form of drowned skyscrapers, stranded ships, empty shopping malls, the rusted bodies of cars, and huge floating islands of debris. Thus, and as this chapter demonstrates, these texts do not retell the familiar saga where white and proficient men from the settler community adapt to the climate emergency to revitalize a militarized yet somehow still benevolent capitalism. Rather, they tell much darker stories of the impossibility of life after capitalism; of the impossibility of life worlds not premised on capitalist relationality.

There are a great number of American climate narratives that evoke such dark futures, and they thus make up an important segment of what has been termed 'climate fiction'. Studied as 'climate fiction' they have been described as texts that introduce the reader to the darkest possible ecological futures and that, through this bleak imagery, demand that readers consider the ecocidal violence 'humans' have performed on the land. However, as the chapter demonstrates, many such post-apocalyptic narratives actually pay homage to the violent settler capitalist system that produced the now crumpled architecture as well as the ruination in which the protagonists find themselves. As the chapter shows, the characters who inhabit these ruined worlds in the texts discussed do not abandon the violent paradigm that informs militarized capitalism. Instead, they keep rehearsing the violent practices that once elevated the US to the position of hegemon within this world-system, and that, as Jairus Grove (2019) and Catherine Besteman (2020) argue, maintain the world-system in the neoliberal present. The violence characters perform in these texts is partly described as a natural response to an ecology that is perceived to have reverted to its endemically hostile, pre-capitalist state. In other words, this future, ruined, and post-capitalist territory is similar to the pre-capitalist geographies that Underhill once securitized through the killing of Indigenous people and the symbolic burning and spoiling of the land, as discussed in Chapter 2. Thus, characters in the narratives studied in this chapter find themselves thrust back into worlds that Thomas Hobbes (1651) once theorized were guided by a 'state of nature' (p. 140) where all beings were involved in a 'warre of every one against every one' (p. 91). In such Hobbesian futures, humans must be violent or they will become victims of violence and thus succumb to post-apocalyptic darkness. At the same time, as this chapter shows, indiscriminate violence is enacted in these narratives also as a kind of capitalist ritual; a celebration of borders and the privatization of property rather than an actual response to the emergency that riotous and untamed nature is described to constitute.

As I have already touched upon, the rehearsal of capitalist tenets in the American Climate Emergency Narrative exemplifies the pervasiveness of what Mark Fisher (2009) has termed 'capitalist realism'. As Fisher makes clear, capitalist realism is premised on the notion that capitalism is the only conceivable social order so that the idea that the capitalist world-system would break down is essentially unimaginable and unreal. If the American Climate Emergency Narrative is premised on such realism, it should be unable to move the action of any story into futures where capitalism is void. However, if the profoundly violent geopolitics through which the capitalist world-system has been realized, and is maintained in the present, is not only a type of global ecological, economic, and social organization but, as Grove has proposed, a 'form of life' (p. 3), capitalist life-making can be imagined to have survived the demise of the system, and to linger as an ontological paradigm, even when it does not remain as a formal social order. In other words, it is possible to imagine futures where capitalism and the capitalist world-system are in ruins, but where the logic, relationships, and ways of being once established by extractive capitalism still structure the storyworld.

The chapter first turns to Nicholas Sansbury Smith's *Hell Divers* (2016), the first novel of a substantial series that describes how an abused and enormously disrupted and disruptive planet has forced what few people that remain to live in a constant state of exception, and where the remnant of a doomed authoritarian, militarized, capitalist state is the only thing keeping humanity from extinction. *Hell Divers* thus illustrates how the American Climate Emergency Narrative invests in the notion of crisis and how it naturalizes the extractive and authoritarian state of emergency as a form of life and the only way of being in the world at a time of full-blown epochal capitalist and ecological crisis. *Hell Divers* sets the stage for the other texts discussed in this chapter: Cormac McCarthy's widely studied *The Road* (2006) and Peter Heller's *The Dog Stars* (2012). These novels take place in the world that *Hell Divers* hints at on every page, but that its plucky soldier heroes constantly postpone. In other words, *The Road* and *The Dog Stars* exemplify the type of story where capitalist society has utterly collapsed, leaving the progeny of white settlers isolated and without the comforts and (in)securities that capitalism furnishes, but where the social forces that made capitalism are still operational.

HELL DIVERS: CAPITALISM AT THE ABYSS

In subaltern writing, the concept of hell is sometimes used as a denominator for the ravaged life worlds created by colonial capitalism. As Andreas Malm (2017) notes, Ghassan Kanafani's novella *Men in the Sun* (1962/ 1999) employs the concept to describe the life of Palestinian refugees escaping the slow violence of the colonial petro-economy. In climate narratives from the core or privileged semiperiphery, where the refugee is a white middle-class subject previously embraced by the comforts of capitalist modernity, the sudden loss of these comforts has also been described as an entry into hell. The German climate breakdown film that narrates a road trip through an impossibly hot and desiccated Germany, where cannibalism is rife and where the final destination of the protagonists provides no relief, is thus succinctly named *Hell* (Fehlbaum 2011).[1] In these different contexts, hell is obviously not the metaphysical, punitive location of Christian, Islamic, or Judean religious texts, but rather an extremely hot (or cold) and permanently insecure space where nationality, gender, race, and class mean little and offer no protection.

In the American Climate Emergency Narrative, author Nicholas Sansbury Smith's *Hell Divers* series, totalling 11 books at the time of writing, employs the concept to name a planet that has collapsed utterly after a third world war. In the series, the land is so polluted by radioactivity and toxic substances, and so ravaged by constant electrical storms, that it has become impossible to live on the surface of the planet. If the electrical storms and lethal radioactivity were not enough, incredibly aggressive life forms are now dominant on the planet's surface. This is a new, resilient, aggressive mutant being intent on killing and consuming its human ancestors wherever they are found. To survive, what remains of humanity has

---

[1] 'Hell' of course means 'bright' in German so that this title refers both to Hell as an imaginary location and to the relentless sunlight that is drying out the world in the film.

been forced to escape in enormous airships that sail above the storms and the fallout that have contaminated the biosphere. To keep the ships from falling back to the planet, and to keep the struggling few hundred on board each ship alive, a cadre known as Hell Divers parachutes through the storms to retrieve fuel cells, pharmaceuticals, and ammunition vital to survival. These soldiers, at constant war with a destroyed and raging planet, are what keeps the remnant of humanity airborne: they 'dive so humanity survives' (p. 7). As in the fiction discussed in Chapter 4, the novel thus externalizes the planet into an essentially hostile and vengeful entity that needs to be constantly fought by what remains of the military tasked with protecting capitalist modernity. This militia is doing the best it can, but things are not going very well. By the time the first novel of the series plays out, only two of the airships remain: the *Hive* and *Ares*. Of these, the *Hive* is dying and *Ares*, in even worse shape, burns and crashes only 100 pages into the text. Humanity is effectively reduced to a gathering of hundreds crammed onto a single and failing airship.

Yet, in this hellish future, life is still organized according to the needs of militarized capitalism. The irradiated and chaotic planet may be inhospitable, guarded by a new, mutant, and utterly hostile indigeneity, but the Earth is clearly also a dysfunctional commodity periphery of sorts since the items needed to keep the airship afloat are sourced from its surface. By extension, the *Hive* is what is left of capitalist society and, like present-day society, life on the airship is clearly stratified. Walking through the ship, Xavier 'X' Rodriguez, a seasoned Hell Diver and the protagonist of the novel, considers how the world he inhabits is organized: 'In some ways, it was even worse now than it had been in the Old World. The caste division of lower-deckers and upper-deckers was painfully apparent everywhere' (p. 78). The 'upper' and 'lower'-deckers are in fact classes rather than casts. The lower-deckers are the precarious, cheap labour of the ship and their food rations and access to health services are often reduced. The comparatively well-off upper-deckers, the Hell Divers and the militia that keeps order on the ship are the privileged social strata, the first to get medical attention and the only ones to eat their fill. A consequence of this stratification is constant social unrest. Clashes between the militia and the lower-deckers are a regular part of life.

This tension feeds the major plot line of the first novel of the series. A man named Travis Eddie has become tired of the constant lack of food, space, and health care. He has lost his (Hell Diver) father and his brother is in jail. With three companions, Travis takes hostages and occupies the part of the ship where animals are raised for food, in the process accidentally damaging the ship. The rebels' demands are straightforward: 'First, you're going to turn the lights and the heat back on in the so-called noncritical facilities. Then you'll double food rations for everyone belowdecks. I want equal health care provided to everyone on this ship. And finally, I want my brother released from jail' (p. 271). These demands may sound mostly reasonable, but the ship's captain cannot meet them even if she wants to. Due to the damage that the rebels have caused, the ship is sinking, and all available energy is needed to keep it from crashing into the storms and radiated ruins below. Thus, lights and heat in the lower decks cannot be turned on.

In this way, the permanent state of emergency in which the people of the airship exist has been aggravated further. This turns the revolution into a selfish and supremely dangerous grab for (undeserved) privileges. The leader Travis may want to bring relief to his lower-deck friends, but his companions are careless and violent. When the ship begins to lose altitude, slowly sinking into the storms that rage below, a situation is produced where the revolution *must* fail or everyone will die. On cue, the brave militia and a plucky teenager manage to save the airship and quell the rebellion. The revolutionaries are put down by precision rifle fire or give up. The airship again rises above the storms and the ruins.

Hell Divers thus describes a future where the conditions that keep militarized capitalism functional have essentially disappeared. The world is an irradiated or ultra-cold post-capitalist ruin. Capitalist social order still exists, but in isolated arks that move through, or hover high above, the ruins that capitalism has produced. This is capitalism on unsustainable life support. If the final ark falls to the ground, capitalism will be effectively over along with the brief reign of humanity. *Hell Divers* is thus premised on the notion that such a descent into the ruins will also be the end of humanity as a species and social form. In other words, the novel builds an imagined future world where the militarized and airborne remnant of capitalist modernity must survive for there to be a social world at all.

The Road: Capitalist Realism in Post-capitalist Futures

As argued, *Hell Divers* tells a story of how an utterly fraught, militarized, capitalist modernity desperately hangs on in futures where life outside the closely bounded world of the airship has reverted into a hyper-charged

Hobbesian war of all against all. Moving beyond this moment of pending collapse, Cormac McCarthy's *The Road* takes place in a future where capitalist modernity has disappeared entirely. Thus, McCarthy's novel can be said to describe the world that awaits people if the airship in *Hell Divers* were to sink to the surface of the planet.

As described in the previous chapter, the first paragraph of this novel introduces the reader to the 'man' and the 'boy' who wake up underneath a plastic tarpaulin, surrounded by an arid, dark, and cold forest. The man knows that they will not survive another winter where they are, so he has stored food and begun a trek to the south in the hope of finding warmer and less hostile climes. The world that they travel through is even more dark, ashen, and ruined than that into which the courageous heroes of *Hell Divers* will descend if the airship fails. All trees are dead or have been burnt to cinders along with most other plant life. There is so much ash and particles in the air that it is necessary to wear (improvised) masks. The odd dog can be heard barking but there are no signs of any other type of animal life. Coming into an abandoned barn, the man realises that cows must now be 'extinct' (p. 101). With no way of growing food for humans or animals, the only sources of sustenance left are what can be scavenged from abandoned houses or stolen from other people. Those less picky turn to other human beings as a food source. 'Bloodcults' and cannibalistic 'marauders' thus haunt the crumbling cities and the road that the man and the boy travel on. So utterly dark are their prospects that the boy's mother has left her family and committed suicide. 'Sooner or later they will catch us and they will kill us' (p. 48), she told her husband before going to her death. 'They will rape me. They'll rape him. They are going to rape us and kill us and eat us' (p. 48).

As the journey towards the south continues, people certainly try to kill the decimated family. Starving, the man and the boy move through the dead landscape, often discovering evidence of cannibalistic meals. In the cellar of a house they break into, they find living humans kept like animals and meant to be consumed piecemeal over time: 'On the mattress lay a man with his legs gone to the hip and the stumps of them blackened and burnt. The smell was hideous' (p. 93). In a camp they enter they come across, a 'charred human infant headless and gutted and blackening' (p. 167) roasting on a spit. The man and the boy flee this scene as they flee all confrontations in the book. The boy wants the man to help some of the people they encounter, to share their food with strangers even more worn and hungry than they are. He has been told by his father that they are the 'good guys' (p. 65), and that they 'carry the flame' (p. 26). But the man is set on his own and his son's survival, on moving on to a never fully realized destination where things will somehow be different. This gives little room for compassion and sharing.

At the end of the story, the man succumbs not to violence or hunger, but to a raking cough that has followed him for most of the journey; the noxious air getting the better of him. In a dreamlike sequence that follows his death and that concludes the novel, the boy is immediately found by a kind and strong man, a 'veteran of old skirmishes' (p. 237), who brings the boy to his loving and welcoming wife and their two small children. Closing with an enigmatic image of 'brook trout' with 'maps of the world and its becoming' (p. 241) on their scaly backs, this final sequence reads more like a weird hallucination or the man's last dying hope for the boy than an event possible within the universe created by the text.

The Road has attracted a considerable amount of scholarship, much of it invested in connecting the novel to McCormac's previous writing, to the legacy of the English literary canon, and to exploring the mythological motifs that are often perceived to inform the text. Thus, in a widely cited article, Lydia Cooper (2011) reads the 'sublimely damaged' world of *The Road* as allegorical, and as an 'apocalyptic grail narrative' (p. 218) similar to T. S. Eliot's *The Waste Land*. Indeed, as Cooper observes, an early draft of the novel was named 'The Grail'. As a grail narrative, then, *The Road* examines, in Cooper's analysis, 'pervasive apocalyptic fears in order to explore if and how the human project may be preserved' (p. 219). Similarly, Carl James Grindley (2008) has argued for understanding the setting of *The Road* as deeply informed by 'Judeo-Christian mythology' (p. 11) and, as such, it is a story about life after the Rapture and the coming of Christ. The enigmatic ending of *The Road* has also been explored in great detail by McCarthy scholars. To Ashley Kunsa, the trout with maps of the world thus 'suggest an inherent order and underlying purpose yet undiscovered' (p. 68).

While McCormac may well have welcomed such analysis of *The Road*, the novel can also be read against this mythologizing grain. When the attention is turned away from religious and spiritual allegory, *The Road* reshapes into a story about post-capitalist and post-climate collapse emergency. Perceived in this very different light, the text communicates a profound sense of epochal ecological and capitalist crisis. *The Road* is littered with artefacts and products that recall capitalism. The man and the boy walk through a blighted America pushing a shopping cart in

front of them. They try to replenish this relic of consumer society when-
ever they can, and it contains everything they possess and all the food
they have to consume. They subside almost exclusively on cans scavenged from hidden-away stores: canned meals and, when they can find
it, soft drinks. In the book and the 2009 film adaptation directed by John
Hillcoat, a can of Coca-Cola is part of a central scene where the boy is
briefly transported back to a pre-apocalyptic era of sugar, aluminium, and
world-system dominance. In this way, as noted by Matt Rainis, the can
of Coke is a 'piece of the ruins of a fallen empire' (2022). As such it
signifies the comforts and privileges that capitalism generates for a certain
stratum of society, but it is equally important as an item once designed to
produce accumulable income from all sections of the world-system. Now
that the world-system has collapsed along with the transportation routes,
the banks, the transactional system, and, indeed, the money that made
up this system, the can of Coke exists only as sugary water and a sign of
capitalist ruin.

The centrality of capitalist ruination has been the focus of a different set
of critical studies of the novel. Simon Schleusener (2017) has argued that
The Road registers capitalism via its 'delineation of a Hobbesian "state of
nature" extrapolated from the basis of our own experience of dog-eat-dog
capitalism' (p. 4). In his reading, the novel 'seems to thoroughly remain
within the confines of the neoliberal imagination' (p. 4). Thus, he argues,
The Road is about capitalism in the sense that it narrates the horrific
experience of being folded into a brutal neoliberal economy where (even
white) middle-class humans are commodities consumed by the system.
The journey through the post-apocalyptic world in which the man and
the boy exist is in fact a passage through the consumptive hell created
by an always hungry, neoliberal capitalism. Read in this way, *The Road*
is fundamentally and aggressively critical of capitalism. It is a novel that,
much like George Romero's *Dawn of the Dead* (1978), shows (undead)
subjects lost in the ruined architecture of consumer capitalism, literally
feeding on each other.

However, *The Road* is arguably just as usefully read as a story about
life in a *post*-capitalist world. In the novel, there are no commodity
frontiers, no class hierarchies, nobody sells their labour and nobody accumulates capital. This is not the class-oriented, primitive capitalist order
that reigns on the airship the *Hive*; it is the world that commences if/
when that contained universe collapses. Schleusener is correct in noting
that the world that the man and the boy inhabit in *The Road* is the one

that Hobbes proposes pre-dates capitalist modernity, where humans are involved in a 'war of all against all', but this world is not identical to the one created through 'dog-eat-dog' capitalism, it is the pre-capitalist universe that 'dog-eat-dog' colonial capitalism reshaped through the establishment of a world-system that made certain people exempt from the violence supposedly inherent in this universe. Again, in the pre-modern state of nature evoked by Hobbes and McCormac, there is no industry, no 'Knowledge of the face of the Earth' and, 'worst of all, continuall feare, and danger of violent death', making 'the life of man solitary, nasty, poore, brutish, and short' (Hobbes 1651, p. 89). According to Hobbes, the modern, sovereign capitalist state (the Leviathan) arrived into this natural world as an engine of a new social order that introduced civilization and prosperity. It is at this moment that property and borders become possible, things and spaces can be owned (and thus stolen), borders tolerated (or transgressed) and social contracts can be formed that regulate human behaviour and allow people to build society. In this way, Hobbes is one of the pillars of early capitalist thinking and he remains foundational to modern-day prophets of neoliberal enlightenment such as Steven Pinker (2011). Importantly, Hobbes' (and Pinker's) claim is not that capitalism/modernity has erased the basic premise by which the brutal natural world functions. On the contrary, the relationality that Hobbes's writing describes is the basic existential condition of life. The natural inclination of all living things is still to engage in a war of all against all. Capitalism has not altered this fundamental state, it has merely created the foundation of a social contract that postpones participation in this war. Should capitalist modernity fail, humanity (narrowly defined) will again be forced to rejoin the strife that, Hobbes and Pinker suggest, is the basic condition of existence.

This is what has occurred in *The Road*. As Carl F. Miller (2014) has observed, *The Road* thus 'articulates a post-capitalist dystopia that offers a mere reversion to pre-capitalism and gives way to general anarchy' (p. 46). In this way, *The Road* is not so much an allegory that reveals the violence inherent in neoliberal capitalism, as a story about what waits on the other side of such capitalism for those white bodies that used to be privileged by it. This makes it possible to understand the 'flame' that the man tells the boy they are keeping as the social contract made possible by capitalist modernity. To carry the flame is to carry a (vain) desire to reignite the material and social order that predated the abject darkness they now walk through. In a world where the extractive word-ecology that creates privilege in the core has ruptured, this privileging social order has become impossible to uphold. In the next instant, *The Road* proposes, children are roasted over open fires.

It is in this way that *The Road* conjures the sense of reality that Fischer has termed capitalist. In this application of capitalist realism, the alternative to capitalism is unthinkable in the sense that the post-capitalist world is narrated as impossibly and absurdly abject and nihilistic: a future reality where there is nothing to eat except ancient preserves and other people so that people *have* to eat other people. By abandoning the reader in such an unbearably cruel post-apocalyptic dystopia, *The Road* enacts a very bleak rehearsal of the capitalist realist notion that there is no valid form of life beyond capitalism. This does not mean that McCarthy is a Pinkerian champion of capitalism. McCarthy's *Blood Meridian* (1985) is clearly devoted to revealing the ruthless violence that accompanied Western capitalist expansion. Yet, as writing from the core, *The Road* is arguably constrained by the perspective that the positionality of its main characters affords. Ultimately, this is a novel that, when it imagines the end of capitalism as the end of the world, comes up with little more than perverse and agonized oblivion. It is possible to argue, as some critics (Dominy 2015; Schleusener, 2017) have done, that *The Road* poses capitalism, or at least humanity, as the reason why the world has collapsed, but even if capitalism is seen to have, in cannibalistic fashion, produced its own demise, *The Road* still describes a functioning capitalist modernity as a much better social world than what follows this modernity's termination. Again, the alternative to the relative safety and affluence afforded by capitalism to some people in the core is an insufferable, ashen world where all people risk sexual violation, cannibalism, and death.

THE DOG STARS: PERFORMING CAPITALIST BORDER-MAKING IN THE RUINS

Since its publication, *The Road* is a text that has haunted the socio-ecological breakdown story in general and the American Climate Emergency Narrative in particular. North American novels such as the aforementioned *Far North* (2009) by Marcel Theroux, *Not a Drop to Drink* (2013) by Mindy McGinnis and the German feature *Hell* (2011), or the more recent, Apple TV+ feature *Finch* (Sapochnik 2021) are some of the many texts that feature landscapes, vulnerable protagonists on the road, and (cannibalistic) marauders who prey on itinerant survivors. Another striking and well-received example of this type of text is Peter Heller's *The Dog Stars*, a novel that opens on a small and isolated airfield somewhere in Colorado where two white men, Hig and Bangley, and a dog named Jasper, are surviving. This is a world transformed by climate change and by a related pandemic officially blamed on Asia but in fact genetically engineered, the reader is told at the end of the novel, in a 'national weapons lab' (p. 253). As a consequence of ecological collapse and the pandemic's 99.9 per cent death rate, capitalist society has crumbled and the modern world has become a ruin. Cities have turned into wastelands full of withered corpses and the few survivors that remain have to walk, like the father and son of *The Road*, through a truly post-apocalyptic, post-capitalist future in search of whatever food and gear that can be sourced from the ruins.

Despite capitalism's demise, however, the novel's first-person narrator Hig is still highly mobile. During the day, he will either hunt with his dog Jasper or take to the skies in an old but still functional Cessna propeller aeroplane. He looks for survivors, potential threats to his small community, and for houses and trucks from which he can scavenge supplies. In the process, he gets an opportunity to temporarily escape the ruins of this world and to remember Melissa, the pregnant wife he lost to the pandemic nine years ago. Meanwhile, Bangley will keep an eye on the fields and forests that surround their refuge. While Hig is something of a failed poet with a potential for empathy, Bangley is a prepper, expert shooter, ruthless killer and a 'Survivor with a capital S' (p. 71). Bangley arrived at the airfield one day loaded with sniper rifles, automatic guns and even a mortar that he saves for a special occasion: a 'surprise sometime, kinda like a birthday present' (p. 135). With the help of this hardware, he keeps the airfield and its occupants safe from intruders.

Bangley does not walk into this novel out of nowhere. In *Saving the Security State* (2017), Inderpal Grewal argues that during what she terms the 'advanced' stage of neoliberalism, citizens have become 'securitized subjects within the United States' (p. 1). This process has generated a set of racialized subjectivities that include what Grewal terms 'exceptional citizens' (p. 1). While these predominantly white figures are often narrated as 'struggling, tragic, or violent' they are also celebrated as 'normative citizen-subjects of the United States as a neoliberal, imperial, security state' (p. 6). Grewal explores four such figures: the 'security mom', the 'humanitarian', the 'security feminist' and, most important for the investigation of *The Dog Stars*, the 'shooter'. This is a 'white male exceptional citizen to whom sovereignty is dispersed so that he can use violence in the protection of the American empire' (p. 6). White, male, and Christian, the shooter 'polices the nation and embodies the white racial sovereignty that he claims to possess' (p. 30).

The fact that the American capitalist empire is effectively dead in *The Dog Stars* does not keep Bangley from embodying the security subjectivity of the shooter as Grewal describes this character. In other words, Bangley is not simply an armed survivor in the post-apocalypse, but a remnant of an otherwise defunct security society. He cannot use his enormous potential for indiscriminate gun violence to maintain the borders of the American empire any longer, but he is able to secure the airfield as if this is the last vestige of a still function fragment of empire. Because Bangley is a shooter and a prepper, such securitization takes the form of pre-emptive violence. On arriving at the airfield, Bangley erected an absolute if invisible border—a perimeter—around an area he considers his and Hig's property. Any human body that crosses this border creates an immediate and overwhelming dearth of security—and emergency—that must be violently managed with the help of indiscriminate gun violence.

The opening of the novel describes one such border violation and how Hig and Bangley respond to it. Hiding in the darkness of the night, Hig spots five men:

> the biggest closest to the dumpster had a rifle with scope, was twisted back doing the talking, signing with his right hand, touching the watch cap cocked on his head, the one just beside him had some sort of assault rifle probably an AK, the three others: two shotguns and a ranch rifle all clear at ninety feet with the goggled eye. The third from the left with a shotgun wore a cowboy hat a short man in a big hat. (p. 44)

Hig is expecting Bangley to do the dirty work, and he eventually does, but not until Hig has put his finger on the trigger and is preparing to strafe the group with automatic fire. At that precise moment, Hig sees the night '[c]racked open [...] the group in my sight coming apart entropic, the red dot flying across like a lethal bug throwing their shadows upwards and out to land, to be swallowed by the green ground' (p. 44). Within seconds, the intruders are dead and Hig and Bangley examine the corpses. Hig has a use for these dead bodies. He is no cannibal, but he turns them into jerky that he feeds his dog. This, the reader is informed, is the dog's favourite food. Hig thus proceeds to turn these dead people into dog food, but he makes one exception when he discovers that the 'short man' he spotted earlier is not a man at all, but 'a boy. Maybe nine. [...] This boy is thin, hair matted and tangled. A hawk feather tied into it. Face hollow, a shadow smirched with dirt and exposure. Would have been born into this. Nine years of this' (p. 46). Hig buries this boy instead of turning him into jerky. He is sorry the child has died, but he still recognizes the necessity of killing walkers. Indeed, most of the people that Hig encounters are truly horrific. One especially monstrous figure he runs into on the road is wearing a necklace of dried vaginas; trophies, it is clear, from a career of extremely violent rape. This man Hig reflexively tears 'open' with his automatic rifle 'Without thought. Left him sprawled back on the road, guts spilled' (p. 89).

These opening scenes explain to the reader what living in the ruins of capitalism means. As in *The Road*, to inhabit this post-capitalist, post-apocalyptic, Hobbesian world is to occupy a space where you must always be ready to kill, or you will yourself be destroyed. As Bangley keeps telling Hig: 'Old rules are done Hig. Went the way of the woodpecker. Gone with the glaciers and the government. New world now. New world new rules. Never ever negotiate' (p. 43). The war of all against all is on again and the only way to survive is to embrace it. The only people that Bangley will tolerate is a small Mennonite community 10 miles south of the airfield, where the children are born with a blood disease that keeps the families from expanding. Bangley will have nothing to do with them, but Hig visits them from time to time, out of compassion.

The novel eventually produces a non-lethal encounter with another group of people. Hig's dog Jasper dies suddenly, depriving Hig of the only companion he truly loves. At the same time, he has intercepted a radio signal from an airfield not too far away. To escape the sense of lone-liness brought on by the dog's death and to investigate the signal, he goes on a long excursion in his plane. On his way, he discovers a secluded valley where a woman and her combative father have hidden from the world along with some surviving farm animals. They practice the same pre-emptive violence as that promoted by Bangley, but the bullets they fire miss Hig and he manages to win their confidence. The woman's name is Cima and she turns out to be Hig's age. Predictably, she is also intel-ligent, inventive, and beautiful. She has been damaged by the pandemic and is fragile, but even so she is the female, heterosexual companion he has been longing for. In addition to this, she is a trained physician and thus another remnant of the now dead security society. Her father is 'long and lean and looked to be strung together with catgut' (p. 206). When inspected more closely, he turns out to be a former special forces soldier, 'Navy SEALs [...] Afghanistan. Other places' (p. 303). In this way, the father is another 'shooter' and thus a man made for this future. When he is told of Bangley's program of indiscriminate long-distance violence, he approves: 'Kinda trained you up. Set a perimeter didn't he? He had no problem killing anything that crossed it. Young, old, men, women' (p. 204).

Hig knows that this is the creed he must continue to live by if he wants to build a community with this new shooter and his canny and beautiful daughter. This is a price he is already paying, and he invites the small family to join him and Bangley at the airport. They load the Cessna up, take to the air and return to the airfield where, it transpires, Bangley has barely survived an attack by a dozen walkers. Cima nurses Bangley back to health and Bangley and her father form an immediate homosocial bond. As shooters, they are the same, ready to do the dirty work necessary to keep their wards alive in an endemically hostile world. With this new and more resilient community in place, they can get on with their post-apocalyptic lives. This has its dark times, but, when Cima's and Hig's love grows and is consummated, also its moments of pleasure.

Sarah E. McFarland has read *The Dog Stars* as an indictment of the 'discourse of progressivism' (p. 3). By imagining a future of extinction where humans have reverted to satisfying basic animal needs, *The Dog Stars* reveals how the human species is always folded into nature and subjected to the same indifferent evolutionary laws as all other animals. In this reading, the novel is an interrogation of 'faulty ideas about human exceptionalism and species difference' (p. 21). Indeed, *The Dog Stars*, like many other fictions set in futures transformed by socio-ecological break-down, is on one level a book about the possibility of human extinction and, as such, also a kind of warning: this is what the world will come to if humans do nothing. By launching this warning, it is very much like *The Road* which takes place in a similar future. In both novels, the lives of the protagonists revolve around the daily grind of survival, almost everyone is out to get you and it makes strange sense to gun children down just to be sure. This is a world without human innocence, where men decorate their necks with the dried vaginas of the women they have raped, murdered, and eaten. In this way, the novel provides no space for thinking that humans are an ordained species, a being set apart from nature and born to dominate it.

However, this reading is premised on an Anthropocene perspective that assumes that humanity, rather than capitalism, has caused the ecological and social collapse the novel describes. When *The Dog Stars* is approached as an Anthropocene text, all human beings are equally guilty of extinction. Even the young boy that Bangley kills and Hig buries rather than feeding him to his dog is somehow to blame. Bangley, Hig, Cima and her father are not innocent either, but they are different in that they have decided not to be like the nomadic killers who fall to their bullets. They are content hunting animals and protecting the enclosure they have claimed for themselves. This somehow turns them into slightly better humans, and it also makes it possible for the reader to like and identify with them. Like the reader located in the hegemonic core of the world-system, they command property of sorts and they are exercising their right to protect it against all introducers, even if the social contract that states that things can be owned and therefore protected is void. Practising such violence is the key to survival in the unorganized and hostile post-capitalist, post-modern Hobbesian wilderness that has replaced the social order once provided by capitalist modernity.

By posing the post-capitalist world as an abject and hostile wilderness where people released from the constraints imposed by the social contract turn into violent rapists and murders, or into survivors forced to pre-emptively massacre these rapists and murderers, the novel's criticism of Anthropocentric climate crisis becomes secondary to its promotion of the type of thinking that has always functioned as a rationale for colonial capitalism. In the writing by Captain John Underhill discussed in Chapter 2, the alternative to the colonial order he works to extend into Indigenous lands is a kind of natural chaos where native agency produces constant emergencies and where property and settlers are constantly insecure. Underhill begs the reader to understand that the genocidal violence he and his men have perpetrated was unavoidable; the only way for them to make the settler community safe. If Underhill's text marks the point in time when capitalism invaded the American continent, Heller's dystopian novel rises out of its imagined demise, and the world that Heller builds around this cataclysmic moment is strangely similar to the one that Underhill narrates. Like the world that Underhill conjures in his writing, this is a space of constant insecurity, where the gun is the only instrument that can keep people safe. Also, just like Underhill, Hig needs absolution for the (necessary) killing he has done, and just like Underhill, he has no intention to stop. The imaginary of capitalism ending thus looks very much like the imaginary that marks its beginning.

What makes this life possible for Hig is not just the presence of a young female and two ready shooters, but the sense that he is upholding the social contract that capitalist modernity established even at a time when this modernity has ceased to exist. In a direct reference to *The Road*, we find out that he feels himself to be 'the keeper of something, not sure what, not the flame, maybe just Jasper' (p. 197). This is false modesty though, because it is exactly the flame that he keeps, the hope of establishing a community where it is at least possible to pretend that the social contract is still valid and operational. Before the demise of Jasper and the arrival of Cima and her father, Hig and his dog were the only ones who could form such a contract (thus, to keep Jasper and to keep the flame were similar endeavours). Once the community is expanded, and under the protective gazes of Bangley and Cima's father, Hig can fan the flames of the old rules into something resembling a fire. He may not be able to establish a true colony or a new commodity frontier to revitalize capitalism, the world is too exhausted for that, but he and his small community can continue to live according to the old rules that state that life within the perimeter is sacred and that all other forms of (human) life can/need to be exterminated with extreme prejudice. In this way, Bangley was incorrect when he stated that the old rules are gone with the woodpecker. The new world is like the old world, only smaller, and it is faith in those old rules that legitimates the killing of all trespassers. Indeed, it is this faith that *turns* itinerant people into trespassers.

Thus, *The Dog Stars* is ultimately a text that *ignores* extinction and that embraces, if with a sad shrug, the extractive and violent relationality that capitalism depends upon. Capitalism lives on as the paradigm that states that territory must be enclosed and privatized, that unenclosed land is profoundly insecure, and that violence is the only way to make it safe again. It is within enclosed and bordered land that the social contract that exorcises the state of nature can be established. For those who have once lived under the rule of this contract, its absence is intolerable. To keep the flame of this contract alive, and to preserve or create the material relationships that make it possible, is foremost on survivors' minds, whether they are walking through the ruined world with only one bullet left in the gun, or if they are comfortably holed up with mortars and a '0.408 CheyTac sniper rifle set up on a platform' (Heller, p. 6). These are the rules and limits that emerge out of the specific register provided by capitalist realism at a time of epochal capitalist crisis.

A Weakness of the Imagination

The American Climate Emergency Narrative is ultimately a body of fiction that gives imaginative shape to the intellectual borders and ontological limits that saturate the core of the capitalist world-system. When read as a world-literature, this narrative can be seen to emerge out of a context where the social injustice that capitalism has produced is not always keenly experienced. Even though the neoliberal gig economy, in its effort to capitalize on increasingly scant Cheap Nature, is producing new anxieties and putting mounting pressure also on the white middle class, capitalism is still typically experienced by this stratum through the comforts it generates: enormous shopping malls, ULED television screens, cars, soft drinks with ice, fast-food hamburgers, online dating services, and so on. The assumption that this is a self-evident material territory, available to all within the world-system who try hard enough, is central to the core culture's inability to envision futures beyond capitalism. In the American Climate Emergency Narrative, this constitutes what Jameson has called a 'weakness of the imagination' (1994, p. xii) that implies that all attempts to preserve capitalism are legitimate and that destroying the planet is preferable to letting it overrun the imaginary borders that separate an externalized and angry nature from a capitalist modernity universalized as human society.[2]

At this particular moment, when what Jason W. Moore (2015) and Giovanni Arrighi (1978, 1994) have described as an epochal capitalist and ecological crisis is unfolding, this weakness of the imagination informs even texts that attempt to tell stories about futures where capitalist modernity *has* collapsed and where people move through the ruins that remain. Pulp action novel franchises such as *Hell Divers* as well as critically acclaimed narratives such as *The Road* and *Dog Stars* thus display this imaginative weakness. These texts' predominantly white and male characters, all born before the apocalypse that has ravaged the biosphere

---

[2] Jameson's original and often paraphrased statement is from his book *Seeds of Time* (1994) and reads: 'It seems to be easier for us today to imagine the thoroughgoing deterioration of the earth and of nature than the breakdown of late capitalism; perhaps that is due to some weakness in our imagination' (xii).

took place, remain attached, like the presumed reader from the core, to a pre-apocalyptic, capitalist world-system that centred on their needs and positionality. Now that this system has collapsed, the world appears as a literal wasteland where their white masculinity means nothing except that there was perhaps more meat on their bones the moment when things went to hell. If capitalism is, as American socialist politician Eugene Victor Debs once suggested 'cannibalism' (1905, p. 18), everyone is suddenly a cannibal or a meal in these narratives' depiction of a Hobbesian, pre/post-capitalist future. The militarized agents of capitalist modernity are no longer there to deal with the emergency that such consumption entails. Emergencies must now be managed by the individual rather than by the security state (a process that is, as Grewal [2017] has shown, already on its way). In *The Road*, the father tries but fails to live up to this role. *The Dog Stars*, by contrast, is peopled by decisive white men who have been preparing for this dearth of security for a long time. In a post-apocalyptic landscape, the shooter comes into his own.

Because the American Climate Emergency Narrative is invested not simply in describing socio-ecological breakdown as an emergency for capitalism, but in leveraging militarized capitalism as the only way to address this emergency, a post-capitalist world must be a world of constant emergency and of perpetual yet ultimately pointless violence. If capitalism is truly dead, the violence can serve no purpose except that of ritual. In this way, curbed by its positionality and the imaginative weakness it confers, the only thing that the American Climate Emergency Narrative set in truly post-capitalist futures can do is to repeat the same violent scenario where those few who keep the flame fend off the mass of humans that have given in to the imperative produced by the return of Hobbes' state of nature. Indeed, *Hell Divers*, *The Dog Stars*, and *The Road* are extremely repetitive texts. The same gruesome scenarios keep occurring over and over. The only way this can end is if those who keep the flame alive all die (as presumably happens in *The Road*). In *Hell Divers*, the story never ends. Every day is an emergency and every new instalment is a lengthening of the franchise. The sense that something is still happening is part of the ploy that capitalist realism uses, but also an illustration of its limitations. Like a magician who performs a single trick over and over again, it keeps telling us the same thing to distract us from the many other stories that could be told, from the idea that something other than a horrendous and bloody state of nature could emerge out of the demise of capitalism.

WORKS CITED

Arrighi, Giovanni. 1978. 'Towards a Theory of Capitalist Crisis.' *New Left Review* 111 (3): 3–24.

———. 1994. *The Long Twentieth Century: Money, Power, and the Origins of Our Times*. London: Verso.

Besteman, Catherine. 2020. *Militarized Global Apartheid*. Durham, NC: Duke University Press.

Cooper, Lydia R. 2011. 'Cormac Mccarthy's "The Road" as Apocalyptic Grail Narrative.' *Studies in the Novel* 43 (2): 218–236.

Debs, Eugene Victor. 1905. *Industrial Unionism*. Chicago: Charles H. Kerr.

Dominy, Jordan J. 2015. 'Cannibalism, Consumerism, and Profanation: Cormac McCarthy's *The Road* and the End of Capitalism.' *The Cormac McCarthy Journal* 13 (1): 143–158.

Edwards, Tim. 2008. 'The End of the Road: Pastoralism and the Post-Apocalyptic Waste Land of Cormac McCarthy's "The Road".' *The Cormac McCarthy Journal* 6: 55–61.

Fehlbaum, Tim, director. 2011. *Hell*. Paramount Pictures.

Fisher, Mark. 2009. *Capitalist Realism: Is There No Alternative?* New York: John Hunt Publishing.

Grewal, Inderpal. 2017. *Saving the Security State: Exceptional Citizens in Twenty-First-Century America*. Durham, NC: Duke University Press.

Grindley, Carl James. 2008. 'The Setting of Mccarthy's *The Road*.' *The Explicator* 67 (1): 11–13.

Grove, Jairus Victor. 2019. *Savage Ecology: War and Geopolitics at the End of the World*. Durham: Duke University Press.

Heller, Peter. 2012. *The Dog Stars*. New York: Alfred A. Knopf.

Hillcoat, John, director. 2009. *The Road*. Dimension Films.

Hobbes, Thomas. 1651. *Leviathan*. 1901. Oxford: Oxford University Press.

Jameson, Fredric. 1994. *The Seeds of Time*. New York: Columbia University Press.

Kunsa, Ashley. 2009. '"Maps of the World in Its Becoming": Post-Apocalyptic Naming in Cormac McCarthy's *The Road*.' *Journal of Modern Literature* 33 (1): 57–74.

Malm, Andreas. 2017. '"This Is the Hell That I Have Heard Of": Some Dialectical Images in Fossil Fuel Fiction.' *Forum for Modern Language Studies* 53 (2): 121–141. https://doi.org/10.1093/fmls/cqw090.

McCarthy, Cormac. 1985. *Blood Meridian: Or the Evening Redness in the West*. New York: Random House.

———. 2006. *The Road*. New York: Alfred A Knopf.

McGinnis, Mindy. 2013. *Not a Drop to Drink*. New York: Katherine Tegen Books.

Miller, Carl F. 2014. 'The Cultural Logic of Post-Capitalism.' In *Blast, Corrupt, Dismantle, Erase: Contemporary North American Dystopian Literature*, edited by Brett Josef Grubisic, Gisèle M. Baxter, and Tara Lee, 45–60. Ontario: Wilfrid Laurier University Press.

Moore, Jason W. 2015. *Capitalism in the Web of Life: Ecology and the Accumulation of Capital*. New York: Verso.

Pinker, Steven. 2011. *The Better Angels of Our Nature: The Decline of Violence in History and Its Causes*. London: Penguin.

Rainis, Matt. 2022. 'Why Viggo Mortensen Fought Fiercely to Get a Can of Coca-Cola into the Road.' *Slash Film*. Accessed June 30, 2023. https://www.slashfilm.com/1114254/why-viggo-mortensen-fought-fiercely-to-get-a-can-of-coca-cola-into-the-road/.

Romero, Geroge, director. 1978. *Dawn of the Dead*. United.

Sansbury Smith, Nicholas. 2016. *Hell Divers*. Ashland: Blackstone Publishing.

Sapochnik, Miguel, director. 2021. *Finch*. Apple TV+.

Schleusener, Simon. 2017. 'The Dialectics of Mobility: Capitalism and Apocalypse in Cormac McCarthy's *The Road*.' *European Journal of American Studies* 12 (12–3). https://journals.openedition.org/ejas/12296.

Theroux, Marcel. 2009. *Far North*. London: Faber and Faber.

CHAPTER 8

Fallout Futures

The Limits of the Capitalist Imagination

In James Tynion IV's and Martin Symmonds' graphic novel series *Department of Truth* (2020, Vol. 1), FBI agent Cole Turner spends his time analysing far-right conspiracy movements, from people who believe that the Earth is flat, to those who subscribe to the theory that the US government has been taken over by shapeshifting reptilians. While attending a Flat Earth convention, Cole is invited by the organizers to attend a cinema screening of the 1969 moon landing that conclusively shows that it was recorded at a film studio by Stanley Kubrick. Cole is then flown in a private airplane to the very end of the Earth where an enormous wall of ice rises up until it meets the sky.

> The Earth *is* flat.
> Turner throws up in the aeroplane toilet.

But the Earth is not, in fact, flat. What Cole has seen, both in the cinema and at the end of the plane ride, are versions of the world generated by the concerted faith of a particular conspiracy movement. The universe within which the story takes place is pliable to such belief. Conspiracy theory is thus capable of warping the actual fabric of reality. After his traumatic visit to the end of the world, Cole is picked up and recruited by the undercover *Department of Truth*. The director of this department explains that 'the more people believe in something, the more true that thing becomes' (2020, no 1. p. 37). This allows the major capitalist actors to actively manipulate the fundamental constitution of the world through the invention and dissemination of outlandish conspiracy theories that stimulate these actors' political and economic agenda. The flat Earth convention and the trip to the end of the Earth have been organized and sponsored by 'Kennet and Bertram Boulet'[1] who have used 'the fortune from their Texas oil company' to fund 'right-wing politicians for the last two decades' (2020, no 1, p. 24). Cole is also told that the conspiracies these and other people like them fund are tied to periodic crises. They are 'a cyclical thing. It happens in times of primal cultural fear' (2020, no 2, p. 27).

By promoting the idea that the capitalist world-ecology is the only possible world and that futures that do not include capitalism are already and always dystopian, the American Climate Emergency Narrative performs work disturbingly similar to that done by conspiracy theories in *The Department of Truth*. These fictions are telling people the world-altering lie that biospheric breakdown is best resolved by investing in the system and strategies that, in fact, have brought the crisis on. Again, this body of texts calls socio-ecological breakdown an emergency, but only to facilitate the employment of the strategies that have always been used to secure extraction: the arming and employment of military and paramilitary units, the building of borders, the creation of surveillance systems, and the dismantling of labour and extraction laws. Even in the case when these narratives attempt to critique the insidious, violent, and profoundly entangled racist, sexist, and cisheteronormative paradigms that have aided extractive capitalism, they fail to offer worlds beyond the horizon imagined by capitalism. The Earth, at the end of the extractive journey, is flat. Capitalism is a great wall of impenetrable ice stretching up into the oblivion of a starry night sky. You cannot move beyond it because there is nothing on the other side.

To expand on this argument, and to summarize the most important points of this book, I have argued that much of what is called climate fiction, defined as a new literary genre profoundly informed by climate science and engaged in an attempt to provide readers with roadmaps towards sustainable futures, is so preoccupied with notions of climate

---

[1] This is a reference to Charles and David Koch, sponsors of a number of anti-science, alt-right groups, including the Flat Earth movement (see McIntyre 2021, 87–88).

emergency and the inevitability of capitalism that a better denominator for many of them is Climate Emergency Narrative. Thus, these texts cast climate breakdown, rather than the processes and systems that have produced it, as the problem. Consequently, the American military security apparatus that has been central to the effort of transforming much of the planet into a resource for extractive capitalism is leveraged as the only system capable of managing this emergency. In the texts studied, socio-ecological breakdown is thus rhetorically resolved through interstate military conflict, through the violent managing of climate refugees, and by combating the planet itself.

The book has furthermore shown how the American Climate Emergency Narrative builds on a long tradition of writing that can be traced, like biospheric breakdown itself, back to the early colonial period. In this way, the American Climate Emergency Narrative emerges out of the same historical and literary tradition, and the same cyclical crises, that have produced the now hegemonic core of the capitalist world-system, or, as Moore terms it, the capitalist world-ecology. In this way, the American Climate Emergency Narrative describes biospheric breakdown from the particular perspective that the hegemonic core constitutes. Again, the world-literature perspective that I have employed—introduced by Franco Moretti (1996) and reworked by the Warwick Research Collective (2015)—poses that the experience of the world-system is combined and uneven: all people across the planet is currently experiencing the same planetary-scale ecological/socioeconomic crisis—the crisis is combined—but they do so in very different ways—the crisis is uneven.

While precarious communities in the peripheries and semiperipheries of the world-system have long been experiencing the violence used to enclose and extract land, the ecological depletion such extractive violence causes, and the deregulatory strategies employed to keep nature and labour *cheap* are only now beginning to affect life in the core of the world-system. This is why this crisis appears as new or as a future emergency in many parts of the Global North. Thus, it is not strange that a socio-ecological crisis that has been unfolding for 400 years is described as new and forthcoming also in narratives from the core. As noted by Kyle P. Whyte (2018) in the article 'Indigenous Science (Fiction) for the Anthropocene: Ancestral Dystopias and Fantasies of Climate Change Crises', there is a prevalent tendency in settler writing to locate climate apocalypse in rapidly approaching futures rather than in the long history that paved the way for the capitalist world-ecology. Similarly, because the US hegemonic core has for such a long invested in, and been privileged by, this world-ecology, it appears, from the perspective of this core, as inevitable and singular: as the thing that saves and the thing that needs saving.

As I have discussed in several of the chapters of the book, many of the texts that I have classified as American Climate Emergency Narratives have been funded and produced by the network of actors tasked with maintaining the core: the US military, the military-industrial complex and the merger between these entities and the entertainment industry that has been termed the Military Entertainment Complex. This is an informal gathering of capitalist institutions and interests tasked with keeping the American nation-state secure and operational. It is not strange that films and other entertainment media generated by this network should struggle to break the bounds of the capitalist imagination. *Indeed, they are tasked with policing imagination as such.* As the book has illustrated, writing from the core not formally supported by the Military Entertainment Complex also tends to toe the line. Even the most dystopian writing from the core, the texts that register the arrival of a socio-ecological breakdown of such proportions that it may be epochal or terminal, struggles to imagine alternatives to capitalism. The solution, in the American Climate Emergency Narrative, is always adaptation and system restoration, not system change. If this fails, darkness falls. The Earth is flat.

FALLOUT

It can be argued that if the climate emergency has been caused by an extractive, militarized capitalist modernity, the American Climate Emergency Narrative is part of its cultural, intellectual fallout. Just like carbon dioxide released by fossil capital, methane produced by industrialized cattle farming, radioactive fallout from nuclear bomb testing, or microplastics discharged from the waste of capitalism, the Climate Emergency Narrative emerges out of the core of the world-ecology to spread across the planet, setting the parameters for how biospheric breakdown can be comprehended and acted upon. It is a narrative that restrains agency by posing that the only meaningful type of action can be performed by and within capitalism, and that argues that radical, anti-capitalist, and anti-colonial action is another engagement opportunity. It is a narrative that asks the reader to have faith in capitalism and to conspire with capitalism against the planet and the people who have long suffered the various crises that the capitalist world-ecology has produced in the peripheries and semiperipheries of the world-system.

This book has been about this fallout and the negative effects it may have also on literary or cultural scholarship interested in ecology and biospheric breakdown. However, by focusing on this type of text, the book also risks helping to cement the very limits built into the climate emergency narrative. If capitalism is not the only future, what other futures are there? The American Climate Emergency Narrative does not provide many clues to this question. This final chapter thus expands the imaginative horizon by considering climate narratives that acknowledge that the borders established between nature and society by capitalism are imaginary, that climate breakdown is not located in the future, but has long been experienced by people in the peripheries and semiperipheries of the world-system, and that capitalism is not an essential or inevitable social order. To this effect, the chapter will consider four different and interconnected types of climate narrative that do not centre capitalism and adaptation. These are *the multispecies, Chthulucene text* that folds humanity into a tentacular natural world, rather than into capitalism, the *Indigenous climate novel* that recognizes how the arrival of extremely violent settler capitalism is the beginning of climate upheaval, what can be termed *the black climate narrative* that centres a climate injustice largely ignored by the American Climate Emergency Narrative from the core, and the *radical, anti-capitalist climate text* that proposes political alternatives to capitalism's seemingly inevitable longevity. The analyses of these texts will be connected to political, sociological, and historical interventions that have voiced strong critiques of the way that dominant capitalist society has remade ecology and human relations.

The narratives discussed in this chapter are also formally American, but they are not fiction from the core. As I discuss, the authors or directors that produce them may live in close proximity to the core, but they have very different relationships to the peripheral and semiperipheral spaces where what Jason W. Moore calls the Four Cheaps (labour, energy, nature, and resources) are extracted and put to work. Stephen Shapiro has defined the semiperipheries as 'the sites where the experience of trauma by peripheral peoples and the speculative entrepreneurship of the core collide to produce new forms of representation, especially as it receives both the oral, folk beliefs of the periphery and the core's printed matter and institutionally consecrated notations, objects, and behavioral performances' (2008, pp. 37–8). This collision enables a different kind of story of the capitalist world-ecology, its possibilities, and futures, and of the strange possibilities that capitalist epochal crisis offers. If fiction from the core is ultimately geared towards celebrating and extending the life world that nurtures the core, fiction from the periphery or semiperiphery is able to query such celebration and extension, and to imagine futures outside of the system that turns the world into core, semiperipheries and peripheries.

Texts that move beyond the imaginary that informs the Climate Emergency Narrative are speculative and future-oriented in ways that the Climate Emergency Narrative is not. As I have argued, the Climate Emergency Narrative pretends that large-scale, socio-ecological crisis is *in the future*, when this is something that has haunted the Global South for a long time. The imaginative leap performed by climate narratives from the peripheries and semiperipheries of the world-system allows it to explore a world truly set in the future: a world where not just the capitalist world-system has folded, but where the extractive logic that drives the system is no longer operational. By doing so, climate narratives from the periphery or semiperiphery often employ what Michael Niblett (2012) and the Warwick Research Collective (2015) term a '"critical irrealist" politics of form' (p. 97). This speculative fictional register shatters the assumptions that guide this narrative and instead employs the apocalypse described in the Climate Emergency Narrative as a door to other, far less dark futures. In other words, if the American Climate Emergency Narrative insists, in ways that recall Fredric Jameson's aforementioned observation that it 'seems to be easier for us today to imagine the thoroughgoing deterioration of the earth and of nature than the breakdown of late capitalism' (1994, p. xii), that the end of capitalism is inevitably the end of everything, the peripheral or semiperipheral climate narrative depicts the end of capitalism as an opportunity for a new type of socio-ecological world. The rest of this concluding chapter probes some of these stories as they refuse to be infected by the intellectual fallout that the Climate Emergency Narrative is spreading across the planet.

Chthulucene Climate Futures

In Canadian-born author Helen Marshall's novel *The Migration* (2019), set in a near future where the climate is eroding in the Global North, a new and inexplicable illness named 'Juvenile Idiopathic Immunodeficiency Syndrome' (p. 34) or JI2 is affecting young people all over the world. Like HIV, it appears to compromise the immune system of the afflicted, making them more susceptible to a range of other illnesses and ultimately killing them. The symptoms are psychological as well as physical, sometimes operating like the parasite Toxoplasma gondii that, as a physician in the book observes, can make those infected more reckless than normal. The novel's protagonist and narrator, seventeen-year-old Sophie Perella, has a sister named Kira who has been diagnosed with this syndrome and they travel with their mother from Toronto to Oxford to receive the best care. This fails to save Kira who succumbs to the illness, leaving her sister and mother in a state of despair. Their sorrow is compounded but also complicated by internet videos that show how people who have died of JI2 experience substantial and disturbing post-mortem tremors. Suspecting that Kira's death may not be final, Sophie wants to save Kira from being cremated. She sneaks into the hospital morgue, finds her sister's body, and drives it to an abandoned factory where she hurriedly abandons it when it begins to shake and quiver.

The Migration makes the connection between the illness and the planetary emergency perfectly clear throughout. Sophie's scientist aunt explains that the Earth System is out of order and that 'the melting of continental glaciers causes massive shifts in already geologically active regions. Storms and flooding here, droughts in Europe and across most of the Middle East. The scale of all this change is unprecedented' (p. 179). The aunt suggests that it is this ecological upheaval that has produced the illness. This also becomes clear when it is discovered that the parts of the world that have been hit the hardest by climate breakdown are also where the illness is most prevalent: 'The areas with the highest mortality rates of JI2 have been China and the Philippines. Bangladesh. It hit earliest in places where there was massive flooding' (179). In this way, the novel importantly acknowledges the uneven nature of climate breakdown. However, what truly sets the novel apart from the conventional climate emergency narrative is its dismantling of the artificial borders that locate humanity as apart from nature, its similar undoing of the notion of regional and nation-state borders, and finally its critique of the state of exception created to manage this illness and its consequences.

Not long after her sister's supposed demise, Sophie finds out that she too has been infected. This makes her abject in the eyes of many people. They leer and shout at her in the street and she will not be able to follow the path set out for her (college followed by a 'productive' life in the Global North). At the same time, she understands that JI2 is not so much an illness as the beginning of a transformative stage. It does not kill young people but sends them into a deep and temporary hibernation during which their bodies acquire a new form. Returning to the factory where Kira's body was abandoned, Sophie collides with her sister who has completed her transformation into a new type of being:

> the slanted shape of bones, the hard keel of her welded ribs, her skin bristling with thousands of tiny pins, a soft layer of flocculent down, new growth. She is trembling, and it's as if she is growing larger, expanding toward me. It's her wings, bursting through the thin membrane of her skin. Those masses on her shoulders, those hulking deformities moving beneath the surface—now stretching out, unfolding. (p. 233)

For lack of a better word, these winged beings are referred to as nymphs and thanks to their wings, they are radically mobile. They can fly for '*thousands* of miles' (p. 255, italics in the original) and thus escape the drowning cities and the parched deserts left behind by capitalist extraction. Just as importantly, they are no longer reliant on the capitalist world-system. Like birds, they travel, find shelter, and feed themselves without its assistance. This means that Sophie is not ill at all. As her physician explains '[y]our body isn't changing because you're sick, Sophie. You manifest as sick because your body has already begun to change, because certain genes have already been activated by the hormone' (pp. 276–7). In other words, the transformation is not an illness, but a kind of evolution. Just like in Nicholas Sansbury Smith's Hell Diver's series discussed in Chapter 7, the planet has altered human biology and produced an alternative, airborne human capable of living outside the capitalist system. However, in absolute contrast to *Hell Divers'* Climate Emergency imaginary, this new type of human species is unaggressive, beautiful and a future of sorts.

In the eyes of the capitalist nation-state and its militarized branches, however, the mobile and independent nature of the new winged humans is extremely problematic. Unable to control them and clearly afraid of the transcendent mobilities they embody, the state responds by creating a state of emergency and an extra-legal strategy of violent repression that allows it to contain and destroy this new, transformative species. Those who appear to die from the infection are cremated to prevent them from completing the change. Those who succumb to the 'illness' in seclusion so that they can complete their transition into the new species are shot down by the police or the military: 'I brought it down', one officer explains on television, 'Three shots, didn't hesitate. It isn't for me to say what it is...but the training, you know, it kicks in right away' (p. 229, ellipsis in the original). The rationale behind the burning and the killing of the (soon to be) transformed is described via Thatcher's capitalist realist adage 'there is no alternative [to neoliberal capitalism]', or TINA.[2] This is voiced by 'Director Ballard' who insists that children are burned before they rise as nymphs and who has told his co-workers that 'there were no alternatives' (p. 280). This is the strategy by which this militarized capitalist modernity intends to adapt to the threat that the new form of independent and vastly mobile human life is imagined to constitute.

Those infected and waiting for the people in power to violently end their new afterlives react with justifiable anger. A friend of Sophie named Reddy hatches a plan to burn a nearby crematorium. Other friends sympathize but also observe: 'You're not a revolutionary, Reddy, you're an English major' (p. 258). But Reddy, evolving, *is* a kind of revolutionary simply because his transformed state allows him to flow through the web of life in a radically different way. Reddy's revolution is expressed partially through his desire to destroy the infrastructure that will be employed to eradicate his future self, but also through his transformation into something that exists entirely outside of the paradigms that capitalism recognizes.

The way that this novel folds humans into ecology by turning them into a kind of border-transcending bird marks a politicized awareness of what Donna Haraway (2016) has termed the Chthulucene. This is a concept that focuses on the entangled and tentacular nature of all life on the planet. To enable a fundamentally extractive relationship to ecology, capitalism needed to sever the lived connections between humans and nature. As scholars such as Bruno Latour (1993) and Val Plumwood

---

[2] See Fisher (2009).

(2005) have observed, and as Moore (2015, 2016) describes in much of his writing, thinkers such as Descartes promoted a dualism where a perceived separation between mind and body came to stand for a separation also between human (as mind) and nature (as body). Posed as an external entity, nature could be thought of as something humans could master, possess, and remake for their own purposes. Because this ontological and epistemological thinking made it possible to pretend that extractive violence did not necessarily affect those (white, male, moneyed people) defined as humans, it was extremely useful to capitalism. As this book has repeatedly argued, this externalizing thinking also informs much writing from the core. Thus, Haraway's concept of the Chthulucene is geared towards making the multispecies connections that capitalism has always attempted to erase from the ontological record visible again. The human body inhabits a multispecies planet. It is dependent on the collaborative work performed by a great number of other species, and it is also host to a multitude of other species.

Marshall's *The Migration* is a novel about climate breakdown that attempts to make readers in the Global North aware first of the fact that humans are as rooted in ecology as the things humans consume. When children and young adults turn into a kind of bird, this centres on how all human beings are in fact animals given life by the planet. This, in turn, makes it possible to understand capitalism's organized extraction of nature as fundamentally hostile also to humans. As evolved, post-capitalist humans, they have also become independent of the borders through which the capitalist world-system has organized the planet. Via the imaginary of a transformed and independent new type of global youth, the violent strategies employed by capitalist modernity to *adapt* to climate change come across as both abject and useless. In *The Migration*, the systematic killing of transformed children and young adults is just as futile as it is horrific; the rehearsal of a bloody fantasy of control. Finally, the suggestion in the novel that the transformed human species will replace the dominant, endemically violent, capitalist modernity that is eroding the planet is fundamentally, if conditionally, utopian. There is a world beyond extractive, militarized capitalist society, even if reaching it requires a profound and foundational transformation of how we think about the human and society.[3]

---

[3] Similar stories are told in novels such as R. M. Carey's *The Girl with All the Gifts* (2014) and *The Boy on the Bridge* (2017), where the metaphor of human multispecies

INDIGENOUS CLIMATE PASTS AND FUTURES

While *The Migration* demonstrates a certain awareness of the uneven nature of socio-ecological breakdown that the American Climate Emergency Narrative is largely oblivious to, it does not recognize the long history of this breakdown and how it has been lived by Indigenous communities for generations. This important history is told by what can be termed the Indigenous, postcolonial, and/or decolonial socio-ecological breakdown narrative that rightly centres on how the arrival of settler capitalism caused what can be termed an emergency for the planet and Indigenous people rather than for capitalism. One of several such texts is Anishinaabe author Waubgeshig Rice's novel *Moon of the Crusted Snow* (2018). This book takes place in the Great Lakes region, on an isolated reservation where people are getting ready for winter by hunting and preparing for what may be a prolonged isolation from the big cities in the south. The novel's protagonist Evan Whitesky makes an effort to observe the traditions of his ancestors and he gives thanks for the moose bull he has just shot. Not everyone on the reservation is as observant. The community has seen its fair share of alcoholism, drug use, and suicides. Even so, it is evidently made up of a resilient group of people who have managed to survive far from the lands in the south from where the Anishinaabe were once deported by white settlers.

On returning to his house, where his wife and two small children are waiting, Evan learns that the satellite television signal has gone down. Mobile phone service, Internet and finally the power also disappears and before long the community understands that something is wrong in the cities to the south. As in any other community, this creates tension. Food and other merchandize sell out from the community's store and no new supplies arrive. The rationing of power, water, and communal food produce a simmering panic. When two young boys who have attended college in the South arrive on snowmobiles, they bring news of a society that is falling apart:

> "It's chaos down there, Izzy," replied Nick. He was referring to Gibson, about three hundred kilometres to the southwest. "The food's all gone. The power's out. There's no gas. There's been no word from Toronto or

---

corporeal transformation opens up sustainable social and ecological relationships beyond capitalism.

anywhere else. People are looting and getting violent. We had to get the fuck out of there". (pp. 74-5)

This is a story told repeatedly in the American Climate Emergency Narrative, but set in an Indigenous community, it plays out very differently. The Anishinaabe community certainly suffers from the lack of fossil-fuel energy and food deliveries from the south, but several hundred years of settler colonialism have made the Anishinaabe into survivors. Under the leadership of elders who have seen worse days, the people on the reservation persist. They will make it through this storm, through the winter, and into a different future.

That is, until the appearance of a white settler on a snowmobile. It is obvious from the beginning that this new arrival is a harbinger of death. Evan sees that '[e]verything was black—the snowmobile, the sled, the boots, the suit, and the helmet' (p. 99), and also that the stranger 'was a beast of a man who was invading his people's space' (p. 100). After declaring that 'I come in peace', the new arrival begins to 'laugh, a mild chuckle, that quickly escalated into sharp guffaws' (p. 100) as if the notion that he may bring peace is enormously comical to him. The name of the newly arrived settler is Justin Scott and he joins up with Evan's low-life brother and other community members who have struggled the most with alcohol and drugs. Justin's authoritarian nature and commandeering personality turn this group away from Evan and the rest of the community. As winter progresses and food becomes more difficult to find, Justin and his hangers-on appear to still eat well. Noticing this, Evan is plagued by strange dreams, and he encounters Justin in one of them:

A tall, gaunt silhouette stood in the doorway, outlined by the scarlet blizzard behind it. The smell made him gag. The creature hunched forward. The hair on its broad shoulders and long arms blurred the lines of its figure. Its legs appeared disfigured, almost backwards. But its large, round head scared him the most. It breathed out another savage rumble.

Evan slowly raised the flashlight, illuminating the figure's pale, heaving emaciated torso under sparse brown body hair. He brought the beam up to its face. It was disfigured yet oddly familiar. Scott. His cheeks and lips were pulled tight against his skull. He breathed heavily through his mouth, with long incisors jutting upward and downward from rows of brown teeth. His eyes were blacked out. If it weren't for the large, bald scalp and the long, pointy nose, this monster would have been largely unrecognizable. (p. 187)

As Gail de Vos (2022) and Priscilla Jolly (2021) have observed, Justin here takes the form of the monstrous, flesh-eating Wendigo that is part of First Nations storytelling. The dream prompts Evan to confront Justin and the crowd he has gathered. He then discovers that Justin and his group have been engaging in cannibalism. Bodies have been stolen from the community's morgue, but Justin has also murdered members of the community and consumed them. In the confrontation that ensues, Justin is shot in the head and dies, but this is not, as in the American Climate Emergency Narrative, a case of extractive violence executed in the name of capitalism. Rather, it is a form of self-defence performed so that settlers will stop eating, and stop encouraging the eating of, Indigenous people.

Through its description of a collapsing white settler society, *Moon of the Crusted Snow* also narrates the arrival of an epochal capitalist crisis, but unlike the American Climate Emergency Narrative, it historicizes this crisis and it also avoids the nihilistic closure of the dominant mode. As the novel unfolds, it traces the present crisis back in time, to its origin in the arrival of extractive settler capitalism. Speaking to the elder Aileen, Evan is told in no uncertain words that:

> Our world isn't ending. It already ended. It ended when the Zhaagnaash [white people] came into our original home down south on that bay and took it from us. That was our world. When the Zhaagnaash cut down all the trees and fished all the fish and forced us out of there, that's when our world ended. They made us come all the way up here. This is not our homeland! [...] Yes, apocalypse. We've had that over and over. But we always survived. We're still here. And we'll still be here, even if the power and the radios don't come back on and we never see any white people ever again. (pp. 149-150)

In this way, to the Anishinaabe community in the novel, the demise of capitalist modernity is not the world-shattering event that the American Climate Emergency Narrative describes it as. Instead, the epochal crisis is a kind of beginning and an opportunity to break out of the violent and extractive system settlers introduced.

Even with this knowledge in mind, and with Justin out of the way, the community is still damaged and divided. As Whyte (2018) notes, there is a tendency even among anti-colonial settler authors to 'privilege themselves as the protagonists who will save Indigenous peoples from colonial violence' (p. 224). The same settler writing furthermore poses Indigenous people as 'Holocene survivors' (p. 236) that the ancestors of white allies 'have not fully harmed through the colonial, capitalist and industrial drivers of the climate crisis' (p. 236) and that, once saved, can reintroduce settlers to a sustainable Holocene economy. *Moon of the Crusted Snow* clearly departs from this narrative through its depiction of the white settler as cannibalistic intruder rather than altruistic saviour prepared to be schooled in (lost) strategies of Holocene survival. To survive the sudden arrival of a distinctly capitalist crisis, this Anishinaabe community must first survive and exorcise the presence of the white settler. With Justin out of the way, the road ahead is still difficult, but not impossible. Thus, the post-capitalist future that the novel envisions is not one where white settlers save Indigenous people from climate death, nor is it the arid, post-capitalist dystopia into which the progeny white settlers are thrown in stories such as *The Road* or *The Dog Stars*. As in *The Migration*, a utopian space opens up at the end of the novel, but this one is not premised on the transformation of an entire species. What is required, from this particular vantage, is simply the eviction of settler capitalism.

A BILLION BLACK ANTHROPOCENES

If the American Climate Emergency Narrative from the core describes capitalist/ecological crisis via the catastrophic loss of comforts and privileges, capitalist colonialism is registered, in the American Indigenous (semi)periphery, through the centuries-long depletion of land, culture, languages, religion, and life that the arrival of white settlers set in motion. Black socio-ecological breakdown narratives emerge out of a similar (semi)peripheral location. Like Indigenous climate fiction, this category of texts is inherently suspicious of capitalism and the extractive violence it performs. It is also a type of text closely related to, if not indistinguishable from, what has been termed Afrofuturism as a genre of writing that uses science fiction to investigate black history and interrogate ongoing social injustice, but also to explore non-dystopian black futures.

One of the trailblazers of this type of writing, and one of the foundational authors of the socio-ecological breakdown narrative, as opposed to the American Climate Emergency Narrative, is Octavia Butler. Her remarkable, prescient, and path-breaking novel *Parable of the Sower* (1993) and its sequel *Parable of the Talents* (1998) tell the story of socio-ecological breakdown in ways that are fundamentally different from the

American Climate Emergency Narrative. *Parable of the Sower* opens on 'Saturday, July 20, 2024' (p. 3) and introduces the reader to a world very similar to those described in many other climate narratives. Thus, this is a future where socio-ecological breakdown has produced resource shortages, increased wealth disparity, and created violent community conflict. Fascism and religious fundamentalism dominate American politics and climate refugees are forced to abandon already unsafe housing to enter roads that are even more dangerous. Unlike in the dominant Climate Emergency Narrative, however, capitalism does little to provide these refugees with futures. The rich secure themselves behind large walls. For those less fortunate, the only alternative is to join one of the many 'company towns' where they will become indentured servants, locked into lives of eternal debt while making a profit for the companies that run these towns.[4]

The protagonist and first-person narrator of the *Parable of the Talents* is Lauren Oya Olamina, a young black woman who turns 15 on the day that the first chapter opens. She lives with her father and her stepmother in a gated and racially mixed community outside of Los Angeles. The people of this community are not rich but are still better off than those who try to survive outside the walls. Led by the father, the people of the community are creating strategies that will help them survive in an increasingly eroding and erosive nation. Things deteriorate further when Christian fundamentalist 'President Donner' is elected and intensifies the already ongoing neoliberal political program introduced to manage the unfolding socio-ecological crisis. By the time Laureen turns eighteen, the social world she is part of is in free fall. Lauren's brother and father become victims of the gangs that run the world outside the gates and a short while later, the borders of the community are breached, and the houses of the families that shelter there are overrun by desperate scavengers.

With her entire family and most of her friends dead, and deprived of all her possessions, Lauren must take to the road. Resourceful and strong

---

[4] As the novel makes clear, these towns are modelled on 'early American company towns in which the companies cheated and abused people' (p. 120). The notion of gated and extractive 'company towns' where the rich find shelter and all other people become precarious labour are used in many other socio-ecological breakdown narratives such as Margareth Atwood's MaddAddam trilogy (2003–2013) and Paulo Bacigalupi's *The Water Knife* (2015).

despite her young years, she gathers up other survivors and begins to trek towards northern California. As in Cormac McCarthy's *The Road* (2006), the journey is extremely hazardous and other travellers take advantage of those who cannot defend themselves. People are killed for their resources, for amusement or as sustenance when there is nothing to eat except other people. Even so, Butler's novel never devolves into a conventional climate emergency narrative. There is no sense that the extractive, neoliberal system that has clearly brought the crisis experienced by Lauren and her fellow travellers on will be able to restore them or the planet. Instead, Lauren works to establish a radically different type of community that she terms Earthseed.

The Parable of the Sower sheds light on, and has also paved the way for, a number of important critical interventions that focus on the connections that exist between socio-ecological breakdown, extractive capitalism, and slavery in America. In *A Billion Black Anthropocenes or None* (2018), Karen Yusoff builds on work by Franz Fanon, Sylvia Wynter, Achille Mbembe, and (speculative fiction author) N. K. Jemesin to problematize what she refers to as the 'structural Whiteness of the Anthropocene' (p. 61). In concert with many other critics of the Anthropocene concept, she thus focuses on the extractivist 'white', 'colonial geo-logics' (p. 2) that systematically elides the constant violence done both to the Earth System (to geology) and to the enslaved people who have been forced to help perform this violence. Her important point is that the Anthropocene concept erases the role that slavery and race have played in this breakdown. In this way, and like Whyte, Yusoff also calls attention to how black (and Indigenous peoples) have already suffered 'a billion black anthropocenes' organized by white extractive capitalism.

As I have argued in this book, the American Climate Emergency Narrative is informed by these precise colonial geo-logics and the structural whiteness that haunts the Anthropocene concept. By contrast, *The Parable of the Sower* clearly works against the logics and structures that inform this dominant narrative. To return to Cormac McCarthy's *The Road* and Brian Hart's *Trouble No Man*, Butler's novel does not centre the plights of white men, but instead casts as the protagonist a young, black woman keenly aware of the long history of extraction, social injustice, and extractive violence through which the capitalist world-ecology was made. Unlike the white protagonists of the American Climate Emergency Narrative, Lauren's life is informed by that historical legacy and the upheaval that she experiences is part of a history that is at the same time racial and personal. Her experience of the destruction of her community is made even more acute by the fact that she suffers from 'hyperempathy syndrome' that forces her to live the pain (and pleasure) of other people as if she experienced it herself. Lauren's hyperempathy thus signifies, as Diana Leong (2016) has argued, black women's 'fleshy existence' (p. 22) as sexualized bodies within the colonial history of the US, but equally her ability to comprehend the suffering of other people.

Butler's novel is furthermore notably different from the dominant Climate Emergency Narrative through its rejection of the capitalist or post-capitalist futures that they project. While the road that Lauren enters is extremely dangerous, it is not the dog-eat-dog, Hobbesian world that dominates the post-apocalyptic stories discussed in previous novels. Lauren encounters dangers but also communities as she makes her way through a crumbling US society. Moreover, if the white post-apocalyptic narrative insists that nothing can change, in the sense that the tenets of capitalism are eternal even in a post-capitalist world, Lauren's Earthseed creed and religion is: 'God is Change' (p. 25). This conviction that change is not only possible, but a divine imperative, drives Lauren's successful attempt to establish a new kind of community. This is no utopia, but it is utopian in its attempt to create a world for 'the living' (p. 3). As such, the formation of the Earthseed community points towards the possibility of a functioning, social world beyond white, Christian fundamentalism and extractive capitalism.

Radical Climate Futures

Alongside Indigenous and black writing from the semiperiphery of the world-system, there is also fiction produced in close proximity to the core that promotes climate futures different from those circulated by the American Climate Emergency Narrative. The most often cited examples of this radical trend in climate fiction are the novels of Kim Stanley Robinson (once the PhD student of Fredrick Jameson). In several science fiction and climate fiction novels, Robinson has built futures that imagine alternatives to the worlds that capitalism is able to conceive. At the time of writing, the most recent example is the novel *The Ministry for the Future* (2020). If 'climate fiction', 'provides us with the beginnings of the roadmap we so sorely need to achieve a global society that is both abundant and sustainable' (np) as Imogen Malpas (2021) has proposed, *The Ministry of the Future* is one of few examples of this genre. Set in a near future where governments are doing 'too little, too late' to prevent global warming, India experiences a week-long heat wave where forest fires and a wet-bulb temperature above 35 kill an estimated 20 million people.[5] This causes the Indian government to spray 'sulfur dioxide' (p. 18) into the atmosphere, this in order to shield the nation from the rays of the sun. The mass death in India also lends authority to an already existing UN organization named The Ministry for the Future, tasked with 'defending all living creatures present and future who cannot speak for themselves, by promoting their legal standing and physical protection' (16). The Indian catastrophe and continued political inertia also provoke the establishment of more or less well-organized eco-terrorist organizations such as the 'Children of Kali' (49) that begin harassing especially erosive international capitalist endeavours, this by downing passenger jets in flight and sabotaging the global meat industry.

The Ministry for the Future is a sprawling, polyphonous novel that considers how the restoration of the Earth System needs to involve a multitude of tactics and solutions. It is narrated via two main protagonists—Mary Murphy, the chairwoman of the Ministry for the Future, and Frank May, an aid worker who has been so traumatized by the unfathomable number of deaths in India that he has turned to violent eco-terrorism—but it is also told with the help of newspaper articles, meeting minutes, poetry, and other types of text. To stop doing harm to the Earth System is not, as Robinson describes it, a straightforward, one-solution endeavour, but an enormously complicated, multi-layered political process that involves the various forces that make up, or resist, the capitalist world-system in very different ways. Meticulously based on up-to-date climate science, the novel explores socio-ecological futures that are reminiscent of, but still very different from, those evoked by the American Climate Emergency Narrative. Like these texts, Robinson's novel includes eco-terrorism, geoengineering, security emergencies, states of exception, green and fossil capitalism, interstate conflict and war, but unlike the dominant narrative, the novel avoids the capitalist realist closure that states that no other future can be imagined other than the restoration of business-as-usual capitalism or devastating post-apocalyptic

---

[5] The wet-bulb temperature is a measurement of the conditions shaped by humidity and temperature combined. At a wet-bulb temperature of 35, the human body cannot regulate its own temperature any longer and will eventually succumb to hyperthermia/ overheating (McMichael, Woodruff and Hales 2006; Dunne, Stouffer and John 2013).

collapse. *The Ministry for the Future* is as dark as it is optimistic, as optimistic as it is dark, yet the ending of the novel reveals a planet where the Earth's temperature is no longer rising, where much of the planet has been left to its own devices, and where socialism operates as a functional alternative to extractive capitalism.

A less often discussed example of the radical climate text is Neill Blomkamp's expensive Hollywood blockbuster *Elysium* (2013). Blomkamp is a South African director best known for his dystopian and anti-apartheid science fiction feature *District 9* (2009), a film that brought him to the attention of Hollywood studio TriStar Pictures. Because *Elysium* is a Hollywood film, there can be little doubt that it emerges out of the capitalist core. It includes product placements for luxury brands Bugatti, and Versace, and for more affordable manufacturers such as Sony and Adidas (Han, 2018). Even so, Blomkamp's ecological and political vision conjures a future beyond capitalism. This vision is often vague and unfocused, but still critically irrealist, rather than unconsciously political.

Elysium describes a future where extractive capitalism has made the planet such a hellish place to live that the privileged have built an enormous space station with its own atmosphere. In white Roman villas surrounded by gigantic green lawns, an almost exclusively white (and French-speaking) population enjoys eternal lives free from disease thanks to 'med bays' that can cure any illness or physical damage in seconds. This is the ultimate gated and militarized community set apart from still Earth-bound people not simply by its elevated location, its armies of formal and informal border guards, but also by a digitized, AI-run political regime that stipulates that only those who live on Elysium are citizens. In Los Angeles (above which Elysium lazily revolves), the better part of humanity eke out utterly precarious lives on the edge of oblivion. To convey what the future of this city might look like once its immensely rich classes have escaped, part of the film was shot in the favelas and enormous garbage dumps of Mexico City (Mirrlees and Pedersen, 2016). By inserting this already existing space of iniquity and poverty into the narrative, the film clearly recognizes the uneven nature of the ongoing socio-ecological crisis.

As in Mexico City and other parts of the world today, inequality is rampant. Almost all people have been deprived of basic dignity and comfort and few of those who can find the money to pay one of the local coyotes to take them on a rogue spaceship to Elysium survive the journey. This is because the ruthless Defence Secretary of Elysium Delacourt will have these ships shot down before they can land on the pristine lawns of the elevated space community. In ways that recall the strategies of the US Department of Defence, she adapts to the challenge that increasing ecological and economic justice entails by violently fortifying the border between the privileged community to which she belongs and the crumbling world below her. At the same time, the film extends the contours of what Mike Davis has termed 'Fortress Los Angeles' (2017) into a future where capitalism is still the roaring engine of a fundamentally uneven world-system.

The film's protagonist Max has always looked towards the gleaming silver wheel that is the space station, working to save up the money needed to pay a smuggler to take him to this promised land. The road to this transition is difficult, however. To find the money needed to travel, Max has stolen cars, spent time in jail and is now on parole. With an ankle monitor strapped around his leg, he travels through slums heavily surveilled by drones and ruthless robots to spend a 12-hour workday at a factory producing the very robots that are used to keep him and his fellow human beings in check. He now has little hope of ever making it to Elysium, and when he is exposed to a deadly dose of radiation during a workplace accident, he is told that his life is over. The owner of the industry, the powerful and absurdly cruel magnate John Carlyle, shows no sympathy. To him, Max is disposable waste or what Giorgio Agamben (1998) has referred to as 'homo sacer': a figure who lacks legal standing and can be killed without consequences. Carlyle thus orders Max to be thrown out of the room where he is recuperating from radiation poisoning because 'I don't want to change the bedding'. This triggers an acute personal crisis fuelled by a wish to live but also by hatred, a crisis that sends Max on a violent journey towards the promised land in space.

Max's new plan is to quickly find the money needed to pay a smuggler to take him to Elysium so that he can use one of the fabled med bays and thus save his own life. But as the film progresses, his project transforms. On board the spaceship that eventually brings him to Elysium are also his former love interest and her acutely ill child. During his preparation for the forbidden transit to the space city, he has also acquired software that makes it possible for him to reset the AI-controlled systems that understand the people of Elysium as citizens while it defines all people of Earth as waste and extractible, subhuman labour. Solidarity with people on Earth, and love for his former girlfriend and her sick daughter, change his agenda. He ultimately abandons the selfish project of restoring his own body for the grander project of revolution. With help from a well-connected crime boss/revolutionary leader, he succeeds in rebooting the system. This allows him to reset the parameters of the system so that *all* human beings become citizens. In the closing scene, the governing AI has dispatched robots with medbays to Earth, thus beginning a dismantling of the capitalist system that turns people to waste.

Blomkamp's political vocabulary, as Ewa Mazierska and Alfredo Suppia (2016) have noted, is clearly Marxist. His film identifies capitalism as the engine of ecological and environmental injustice and his white working-class hero (played by Matt Damon) decides to sacrifice his life to complete the violent but still noticeably unbloody and algorithmic revolution that takes place in the film. It is not a perfect film by any standard. As observed by Christa Van Raalte (2015) it celebrates the white, hypermasculine hero and it pays only marginal attention to the sexist, heteropatriarchal, and racist epistemologies that enable and structure capitalism. Also, it says very little about what type of society will emerge now that all people on Earth are citizens of Elysium. Will the sharing of Elysium's vast resources and technology produce an egalitarian utopia or will all people now be part of the capitalist world Elysium is the crowning achievement of? Even with these questions in mind, *Elysium* is trying to say something very different from the dominant climate emergency narrative. In this Hollywood film, the emergency is obviously the suffering that socio-ecological breakdown is causing, aggravated by the state of exception declared by the capitalist security state. In other words, Elysium takes place in a storyworld where the biospheric crisis is one element of an interrelated metacrisis that deprives all but the one per cent (in the film soaring above the Earth in the ultimate gated community) of security, food, dignity, and hope.

Rebooting the System

While these four categories of climate writing focus on different histories and experiences, and propose different developments and futures, they are not separated by fixed boundaries. They intersect at times, they stress similar organic relationships to the planet, and they all reveal the emergency to be the violence done to the planet, to multispecies ecologies, and to the Indigenous, racialized, and impoverished people that are leveraged as resources or waste rather than treated as humans. When capitalist colonialism is perceived as the engine of emergency, rather than its only antidote, it becomes possible to imagine futures where it does not reign. These futures are not necessarily utopian, but they are not post-capitalist wastelands either. It is possible to imagine the end of capitalism and the beginning of another set of relationships that do not understand the planet and other people as resources to be extracted.

Such counter-narratives are important at this point in time. The story told about socio-ecological breakdown by the American Climate Emergency Narrative arguably dominates not only fiction, but also the conversation that takes place outside of the cultural sphere. Few governments in the Global North are involved in an organized effort to save the planet or those who have long suffered from socio-ecological erosion. Rather, most such governments are feverishly trying to save capitalism. Long before the fall of the Berlin Wall and the Soviet Union, and Francis Fukuyama's (1992) declaration that 'we' are now at the end of history, the dominant political vision has been shaped by liberal/colonial capitalism. The shifting positions of Russia, China, and India within the world-system may be causing great concern in what is still often called the West, but this has not altered 'Western' governments' faith in capitalism as the only world order. However, as tensions rise between these major actors of the world-system, as 'cheap' natural resources are increasingly depleted, climate refugees grow in number and nation-states and alliances invest more and more of their proceeds in the military,[6] it is becoming increasingly difficult to believe that the capitalist world-system will produce a sustainable, politically sound, and socially just future.

Because of this, it is necessary to shift the conversation. As Evan Calder Williams (2011, p. 5) has argued, what is needed is 'an apocalypse'; not one of '*total destruction* but rather a destruction of *totalizing* structures, of those universal notions that do not just describe "how things are" but serve to prescribe and insist that "this is how things must be"' (p. 2). What form can such apocalypse take? What futures can take place beyond the capitalist realist insistence that the future is either capitalist regeneration or the resurrection of Hobbes' war of all against all? Hannah Holleman is one of many who has proposed a radical program for Earth-system survival and human social justice that does not build on capitalist

---

[6] As described by the Stockholm International Peace Research Institute's publication 'Trends in World Military Expenditure' (Tian et al. 2023), global investment in the military grew by 3.7 per cent in 2022, to the record sum of $2240 billion.

renewal. In *Dust Bowls of Empire* (2018), she dismisses mainstream environmentalism's strategy to alleviate climate breakdown with the help of Green Capitalism or via the three Rs: 'reduce, reuse, recycle' (p. 162). While these serve a purpose, they are not what will restore the planet. Such restoration instead needs to be performed via a different set of Rs: 'restitution (of lands and sovereignty, of power to the people), reparations (for slavery, stolen labor, genocide, and other past injustices), restoration (of earth systems), and revolution (moving away from capitalism)' (p. 162). In this way, Holleman argues that Earth-system restoration must begin with the establishment of a robust and socially just system that does not work according to the logics of extractive capitalism.

When propositions such as these are not dismissed by liberal thinkers with a flick of the wrist—as already proven wrong, as absurdly utopian, as catering for the needs of a small minority, or as averse to 'human nature'—its proponents are being harassed or even murdered and their activism and scholarship outlawed.[7] Yet, and to emphasize a point already made, the observation that the socially and economically unjust world created by the capitalist world-system is driving the ecological crisis is now also being made by climate scientists who do not emerge out of decolonial, feminist, or Marxist intellectual traditions. The 2022 IPCC report (Lee et al. 2023) argues that '[a]ctions that prioritise equity, climate justice, social justice and inclusion lead to more sustainable outcomes, co-benefits, reduce trade-offs, support transformative change and advance climate resilient development' (66). Similarly, the aforementioned *Earth for All* (Dixson-Declève et al. 2022) combines hard climate science with computer simulations that reveal, as Christiana Figueres (2022) observes, that 'the climate crisis, the nature crisis, the inequality crisis, the food crisis all share the same deep root: extractivism based on extrinsic principles' (p. xvii). It follows that to come to terms with biospheric breakdown, it is necessary to address these related human and ecological crises in unison. To this effect, *Earth for All* proposes five turnarounds: (1) Ending poverty, (2) Addressing gross inequality, (3) Empowering women, (4) Making our food system healthy for people and ecosystems, and (5) Transitioning to clean energy (p. 5). In this way, this publication promotes an

---

[7] See, for example, Smith (2008), Watts (2016), Watts and Vidal (2017), Jeffords (2016), Lynch et al. (2018), Glazebrook and Opoku (2018), and Middeldorp and Le Billon (2019).

'essential reboot of civilization's guiding rules before the system crashes' (p. 29).

Such visions for the future launched by the authors of the reports issued by the Intergovernmental Panel on Climate Change resist the paradigm rehearsed by the American Climate Emergency Narrative. As this book has demonstrated, this conservative type of text is premised on the Anthropocene model of the world where all humans have somehow contributed to biospheric breakdown, and it focuses on the planet and the conflict that Earth-system breakdown will produce as an opportunity for capitalism to save (a certain segment of) humanity by going to battle; by making war on the planet, by engaging other states or by combatting the displaced millions that are forced onto the road by floods and droughts. It is a narrative written from the core of the capitalist world-system, and while it registers the arrival of an epochal crisis, it insists that the emergency that such crisis constitutes must be addressed by the system, and must serve to preserve the system. In the few instances where a world after capitalism is envisioned, it takes the shape of an abject darkness of such proportions that it resembles Hell.

Being aware of the limitations of this type of text is essential when considering what has brought on the current crisis, and when thinking about how to address it. Biospheric breakdown did not arrive like an asteroid from outer space or as the inevitable result of human evolution. It has a material history that begins with capitalist colonialism, and it is an effect of the social, ecological, and economic injustice caused by this system. As 'we' and our governments prepare to act upon this crisis, awareness of this history and its intellectual consequences—of the climate emergency narrative as a form of fallout—is essential. Biospheric breakdown must not be thought of as a military engagement opportunity but as an impetus to reorganize the very system that is destroying people and ecologies and calling it progress.

The world is not flat and there is no impenetrable wall at the end of the journey. The trick is to imagine a movement forward and through and a life beyond based on other values.

WORKS CITED

Agamben, Giorgio. 1998. *Homo Sacer: Sovereign Power and Bare Life*. Chicago: University of Chicago Press.
Atwood, Margaret. 2003. *Oryx and Crake*. Toronto: McClelland & Stewart.

————. 2009. *The Year of the Flood*. Toronto: McClelland & Stewart.

————. 2013. *MaddAddam*. Toronto: McClelland & Stewart.

Blomkamp, Neill (dir.). 2009. *District 9*. Sony Pictures.

———— (dir.). 2013. *Elysium*. Sony Pictures.

Butler, Octavia. 1993. *Parable of the Sower*. 2023. New York: Grand Central Publishing.

————. 1998. *Parable of the Talents*. 2023. New York: Grand Central Publishing.

Carey, M.R. 2014. *The Girl with All the Gifts*. London: Orbit.

————. 2017. *The Boy on the Bridge*. London: Orbit.

Davis, Mike. 2017. 'Fortress Los Angeles: The Militarization of Urban Space'. In *Cultural Criminology*, edited by Keith Hayward, 287–314. Abingdon: Routledge.

Dixson-Declève, Sandrine, Owen Gaffney, Jayati Ghosh, Jørgen. Randers, Johan Rockström, and Per Espen Stocknes. 2022. *Earth for All: A Survival Guide for Humanity: A Report to the Club of Rome*. Gabriola Island: New Society Publishers.

Dunne, John P., Ronald J. Stouffer, and Jasmin G. John. 2013. 'Reductions in Labour Capacity from Heat Stress under Climate Warming.' *Nature Climate Change* 3 (6): 563–566.

Fisher, Mark. 2009. *Capitalist Realism: Is There No Alternative?* New York: John Hunt Publishing.

Fukuyama, Francis. 1992. *The End of History and the Last Man*. New York: Free Press.

Glazebrook, Trish, and Emmanuela Opoku. 2018. 'Defending the Defenders: Environmental Protectors, Climate Change and Human Rights.' *Ethics and the Environment* 23 (2): 83–109.

Han, Angie. 2018. 'How Neill Blomkamp Chose the Product Placement in "Elysium".' *Slashfilm*, June 18. https://www.slashfilm.com/526741/how-neill-blomkamp-chose-the-product-placement-in-elysium/.

Haraway, Donna J. 2016. *Staying with the Trouble: Making Kin in the Chthulucene*. Durham, NC: Duke University Press.

Holleman, Hannah. 2018. *Dust Bowls of Empire: Imperialism, Environmental Politics, and the Injustice of "Green" Capitalism*. New Haven, CT: Yale University Press.

Jameson, Frederic. 1994. *The Seeds of Time*. New York: Columbia University Press.

Jeffords, Chris, and Alexa Thompson. 2016. 'An Empirical Analysis of Fatal Crimes against Environmental and Land Activists.' *Economics Bulletin* 36 (2): 827–842.

Jolly, Priscilla. 2021. 'Thinking About the End of World with Kathryn Yusoff and Waubgeshig Rice.' *The Goose* 19 (1): art 3.

Latour, Bruno. 1993. *We Have Never Been Modern*. Cambridge, MA: Harvard University Press.

Lee, Hoesung, K. Calvin, D. Dasgupta, G. Krinner, A. Mukherji, and P. Thorne. 2023. 'Synthesis Report of the IPCC Sixth Assessment Report (Ar6).' *Intergovernmental Panel on Climate Change*. Geneva, Switzerland.

Leong, Diana. 2016. 'The Mattering of Black Lives: Octavia Butler's Hyperempathy and the Promise of the New Materialisms.' *Catalyst: Feminism, Theory, Technoscience* 2 (2): 1–35.

Lynch, Michael J., Paul B. Stretesky, and Michael A. Long. 2018. 'Green Criminology and Native Peoples: The Treadmill of Production and the Killing of Indigenous Environmental Activists.' *Theoretical Criminology* 22 (3): 318–341.

Malpas, Imogen. 2021. 'Climate Fiction Is a Vital Tool for Producing Better Planetary Futures.' *The Lancet Planetary Health* 5 (1): e12–e13.

Marshall, Helen. 2019. *The Migration*. London: Titan Books.

Mazierska, Ewa, and Alfredo Suppia. 2016. 'Capitalism and Wasted Lives in *District 9* and *Elysium*.' In *Red Alert: Marxist Approaches to Science Fiction Cinema*, edited by Ewa Mazierska and Alfredo Suppia: 121–148. Detroit, Wayne State University Press.

McCarthy, Cormac. 2006. *The Road*. New York: Alfred A Knopf.

McIntyre, Lee. 2021. *How to Talk to a Science Denier: Conversations with Flat Earthers, Climate Deniers, and Others Who Defy Reason*. Boston, MA: MIT Press.

McMichael, Anthony J., Rosalie E. Woodruff, and Simon Hales. 2006. 'Climate Change and Human Health: Present and Future Risks.' *The Lancet* 367 (9513): 859–869.

Middeldorp, Nick, and Philippe Le Billon. 2019. 'Deadly Environmental Governance: Authoritarianism, Eco-Populism, and the Repression of Environmental and Land Defenders.' *Annals of the American Association of Geographers* 109 (2): 324–337.

Mirrlees, Tanner, and Isabel Pedersen. 2016. 'Elysium as a Critical Dystopia.' *International Journal of Media & Cultural Politics* 12 (3): 305–322.

Moore, Jason W. 2015. *Capitalism in the Web of Life: Ecology and the Accumulation of Capital*. New York: Verso.

———. 2016. *Anthropocene or Capitalocene? Nature, History, and the Crisis of Capitalism*. Oakland: PM Press.

Moretti, Franco. 1996. *Modern Epic: The World System from Goethe to García Márquez*. London: Verso.

Niblett, Michael. 2012. 'World-Economy, World-Ecology, World Literature.' *Green Letters* 16 (1): 15–30.

Plumwood, Val. 2005. *Environmental Culture: The Ecological Crisis of Reason*. Abingdon: Routledge.

Rice, Waubgeshig. 2018. *Moon of the Crusted Snow: A Novel*. Toronto: ECW Press.

Robinson, Kim Stanley. 2020. *The Ministry for the Future*. London: Little, Brown Book Group.

Shapiro, Stephen. 2008. *The Culture and Commerce of the Early American Novel: Reading the Atlantic World-System*. University Park: Penn State University Press.

Smith, Rebecca K. 2008. 'Ecoterrorism: A Critical Analysis of the Vilification of Radical Environmental Activists as Terrorists.' *Environmental Law* 38 (2): 537–576.

Tian, Nan, Diego Lopes da Silva, Xiao Liang, Lorenzo Scarazzato, Lucie Béraud-Sudreau, and Ana Carolina de Oliveira Assis. 2023. *Trends in World Military Expenditure, 2022*. Stockholm: Stockholm International Peace Research Institute.

Tynion, James IV, and Martin Symmonds. 2020. *Department of Truth*, vol. 1. Portland: Image Comix.

Raalte, Van, and Christa. 2015. 'No Small-Talk in Paradise: Why Elysium Fails the Bechdel Test, and Why We Should Care.' In *Media, Margins and Popular Culture*, edited by Einar Thorsen, Heather Savigny, Jenny Alexander, and Daniel Jackson, 15–27. Basingstoke: Palgrave Macmillan.

de Vos, Gail. 2022. 'The Windigo as Monster.' In *North American Monsters: A Contemporary Legend Casebook*, edited by David J. Puglia, 282–297. Louisville, CO: Utah State University Press.

Watts, Jonathan. 2016. 'Berta Cáceres, Honduran Human Rights and Environment Activist, Murdered.' *The Guardian*, 4 March. https://www.the guardian.com/world/2016/mar/03/honduras-berta-caceres-murder-enivro nment-activist-human-rights

Watts, Jonathan, and John Vidal. 2017. 'Environmental Defenders Being Killed in Record Numbers Globally New Research Reveals.' *Chain Reaction* 130: 40–41.

Warwick Research Collective (WReC). 2015. *Combined and Uneven Development: Towards a New Theory of World-Literature*. Liverpool: Liverpool University Press.

Whyte, Kyle P. 2018. 'Indigenous Science (Fiction) for the Anthropocene: Ancestral Dystopias and Fantasies of Climate Change Crises.' *Environment and Planning E: Nature and Space* 1 (1–2): 224–242.

Williams, Evan Calder. 2011. *Combined and Uneven Apocalypse: Luciferian Marxism*. Hants: Zero Books.

Yusoff, Kathryn. 2018. *A Billion Black Anthropocenes or None*. Minneapolis: University of Minnesota Press.

INDEX